FOOD SCIENCE AND TECHNOLOGY

BIOACTIVE COMPOUNDS AND THEIR IMPORTANCE

FOOD SCIENCE AND TECHNOLOGY

Additional books and e-books in this series can be found on Nova's website under the Series tab.

FOOD SCIENCE AND TECHNOLOGY

BIOACTIVE COMPOUNDS AND THEIR IMPORTANCE

GULCIN YILDIZ

AND

GÖKÇEN İZLİ

EDITORS

nova
science publishers
New York

DOI: https://doi.org/10.52305/AFDL7660

NOTICE TO THE READER

Library of Congress Cataloging-in-Publication Data

ISBN: 978-1-68507-361-9

Published by Nova Science Publishers, Inc. † New York

Contents

PREFACE

This book includes eleven chapters about bioactive compounds and their importance. Chapter One describes the effects of drying application on bioactive compounds of fruits and vegetables. Chapter Two analyzes the effect of different drying degrees on antioxidant capacity and total phenolic content of red capia pepper. Chapter Three presents data related to the encapsulation of bioactive compounds from herbal extracts using spray drying. Chapter Four summarizes the impact of bioactive compounds on animals and the environment. Chapter Five discusses the bioactive compounds in plants and their wound healing properties. Chapter Six details the biodiversity and health effects of sesquiterpenes in aromatic and medicinal plants. Chapter Seven highlights phytochemicals and their bioactivity for humans. Chapter Eight describes the information on bioactive polysaccharides in tropical fruits and their relationship with potential beneficial health effects. Chapter Nine provides the significance of marine bioactive natural products and their applications as emerging drugs. Chapter Ten summarizes the enrichment of eggs and meat with bioactive compounds. Finally, Chapter Eleven provides a critical view about a valuable bioactive molecule called Lycopene.

Chapter 1 - Many bioactive compounds consist of antioxidants, vitamins, carotenoids, flavonoids, and other phenolics that exist in fruits and vegetables in appreciable amounts. Bioactive compounds have different biological activities such as scavenging reactive oxygen species (ROS), inhibiting oxidation and microbial growth, and reducing the risk of chronic diseases such as cancers and cardiovascular disorders. However, they have low stability against environmental conditions. Therefore, the drying process emerged to decrease their sensitivity and present a target delivery system. Dried fruits and vegetables can be easily stored and transported at a relatively low cost. They have reduced packing costs, and their low moisture content delays microbial spoilage. This review provides an overview of recent findings on the effects of different drying techniques on major bioactive compounds of fruits and vegetables. In particular, changes in ascorbic acids, carotenoids, flavonoids, phenolic acids, total phenolics, and antioxidant capacities are discussed in detail.

Chapter 2 - This work was undertaken to evaluate the effects of different drying temperatures (50, 60, and 70 °C) and cutting types (ring and cube cutting) on the total phenolic content and antioxidant capacity of red capia pepper during storage (14 days) both at room temperature and cold storage. Convective drying (CVD) at 50, 60, and 70 °C was applied to get dried red capia peppers. The effects of CVD on selected bioactive compounds (total phenolic content and antioxidant capacity) of thirteen dried red capia peppers were compared. In addition, the effects of the cutting types (ring and cube) on the final quality of pepper samples were determined both at room temperature and cold storage during two weeks (14 days). The pepper samples dried at 60 °C in ring forms (6R-CS samples) resulted in the highest antioxidant capacity and total phenolic content of all storage times. Overall, dried red capia peppers in ring forms at 60 °C can be used to acquire a high-quality food product with an enhanced total phenolic content and antioxidant capacity.

Chapter 3 - In the past few decades, the spray drying technique has gained wide acceptance as an alternative drying technique for transferring liquid forms into stable dry powder forms with application in the food, pharmaceuticals and cosmetic industry. Spray drying technique enables short operation time, drying of heat sensitive materials, desirable moisture content and desirable particle size of obtained powder. For the centuries, medicinal plants are considered as one of the most important sources of bioactive compounds. Different plant parts including leaves, flowers, stalks, fruits, seeds and roots but also whole plants are used for producing herbal pharmaceuticals, herbal cosmetic and dietary supplements. One of the most common used form are herbal extracts, produced with different extraction techniques (conventional and green novel techniques). Herbal extracts are highly concentrated in bioactive compounds such as phenolic compounds, tannins, alkaloids, pigments, vitamins, bioactive polysaccharides, peptides and volatile compounds with different biological properties. Liquid form of herbal extracts suffers from many disadvantages, such as short shelf life, higher cost of transport and storage. This chapter provides an overview on the principal applications related to encapsulation of bioactive compounds from herbal extracts with special attention on the choice of encapsulation agent on the quality of final dry extract. The application of spray drying represents a new promising drying technique for the production of dry extracts rich in bioactive compounds, which can be easily incorporated into new highly-valuable, bioactive compounds-rich products.

Chapter 4 - The use of alternative feedstuffs in animal nutrition has been associated with several benefits regarding the environment, the quality of the derived animal products, and the food-feed competition. Such potential feed sources include oilseeds and their by-products seaweeds and algae, food waste and agro-industrial by-products containing a plethora of bioactive compounds, namely long-chain polyunsaturated fatty acids, essential oils, polyphenols, flavonoids, carotenoids, and other antioxidants and immunoregulating compounds. Addition to animals' diets may result in producing functional animal products that meet the global market trends and fortify both humans' and animals' health. The present chapter will focus on the role of bioactive compounds present on agro-industrial by-products, innovative feeds and biotechnologically processed derivates on animal performance, the quality of the derived animal products (eggs, meat, and milk), the livestock health status, and the associated environmental benefits.

Chapter 5 - Wound healing is an intricate and active process demanding a highly regulated cascade of biochemical and cellular events designed to restore tissue integrity following injury. Studies have revealed that nature provides a vast number of plants and other products with significant wound healing activities. Several plants have the ability to stimulate the process of wound healing. Plant extracts have different compounds that play an essential role in healing because of their natural repair mechanism. Their extracts have been found to have a profound potential in managing and treating wounds. They are reported to be safe and hypersensitive reactions rarely occur with the use of these agents. Most plants have wound healing activity due to the synergistic or additive actions of their constituents. Plant parts commonly used for wound healing are leaves followed by roots; each plant, the part's therapeutic property varies. Usually, the leaves' fresh pastes are made from the plant and are applied topically for practical use. The plant parts' medicinal property goes so that different plant parts are used in various forms, based on preparation. Modern medicine plays a significant role in treating wounds, but the number of medications is limited. Medicinal plants are frequently used as raw materials to extract active ingredients used to synthesize different drugs. Although traditional medicines offer a safe, inexpensive approach to treating wounds and burns, it has not received adequate attention. In this study, the authors collected the data from various Ayurvedic books and consulted with eminent doctors of Ayurveda and Modern medicine. By analyzing the parameters, selected plants can be further classified based on the type of disease to which they are used. The present study aims at disseminating the knowledge of traditional flora using discrete formulations treated for wound healing.

Chapter 6 - The importance of natural products in the treatment of various health conditions, as well as in drug discovery, is impressive. This is supported by the fact that over three quarters of the world's population mainly relies on plants and herbal products for health care. According to the study provided by World Health Organization in 2002, more than 30% of plant species in the world were used for medicinal purposes. The presence of various bioactive compounds is the main reason for the widespread use of aromatic and medicinal herbs, both in folk and modern medicine. These bioactive components, showing positive health effects in various health conditions, are the focus of numerous studies.

Terpenes, a group of highly bioactive plant components, play an important role in aromatic plants considering the ecological chemistry of interaction with insects and pathogens. Terpenes are one of the main constituents of the plant essential oils, most commonly used in the flavor and fragrance industry. Their presence in the plants is highly dependent on the plant species and growing conditions. Many essential oil terpenes possess significant positive health effects and pharmaceutical importance, which encourages further research on their mode of formation, characterization and biological activity.

Terpenes are classified according to their skeletal structure, with the most common being monoterpenes, diterpenes and sesquiterpenes listed as a group of 15-carbon isoprenoid compounds derived from the assembly of three isoprenoid units. In higher plants, sesquiterpenes are represented by several acyclic, mono, bi-, tri- and tetracyclic systems, many of which possess significant biological activity such as antimicrobial, antitumor, and cytotoxic properties. Thousands of sesquiterpenes known today are formed from 300 different C15-hydrocarbon skeletons that are produced from the substrate farnesyl diphosphate (FPP) by the action of sesquiterpene synthases.

This chapter highlights the biodiversity of plant sesquiterpenoids as well as their health effects, focusing mainly on modern health problems. The appearance of different types of sesquiterpenes as well as a comparison of their structure and biological activity, with the emphasis on anticancer activity, is reviewed.

Chapter 7 - Plants are sessile organisms that evolved several mechanisms to cope with the adversities imposed by the environment. One of the strategies to deal with environmental stressors is the production of compounds known as specialized/secondary metabolites. These compounds are known for not being present in all plants and not being directly involved in the growth, development, or reproduction of the organism. In addition to stressful situations, plants also accumulate specialized metabolites in physiological processes, such as fruit ripening. These compounds are also important to fundamental plant processes, such as photosynthesis, respiration, and phenotypic plasticity. However, the human intellect learned to use these substances for other purposes not related at all to plant survival and maintenance. The authors take advantage of the potential of these compounds and use them to fight or prevent problems associated with several diseases, such as stress. The main goal of this chapter is to link the importance of the phytochemicals in the plant to the bioactive compounds in the human diet. Phenolic compounds, carotenoids, ascorbic acid, alkaloids, glucosinolates, among others, will be discussed. The authors believe that a broad knowledge of the function of phytochemicals is of the utmost importance for students, professors, and researchers to better understand the main concerns about bioactive compounds. The use of moderate stress techniques to improve the quality of plants in terms of bioactive compounds, as well as studies related to extraction and optimization techniques, food biofortification and enrichment, encapsulation, and nanotechnology for application in food, molecular engineering, plant tissue culture, and studies with unconventional food plants are examples of the importance of cooperation and knowledge from different areas of science. Despite the large number of studies already conducted on phytochemicals and their bioactive capacity, there is still much to be explored.

Chapter 8 - Tropical fruits grow in tropical or subtropical climates, in the geographical area extending from latitude 30 ° south to latitude 30 ° north. The temperature conditions in this area vary between 16 to 36 ° C during the year. These fruits are important to developing countries from a nutritional and bioactive point of view. Their pulp, peel and seeds are rich sources of bioactive compounds such as phenolics acids, polyphenols, betalains, carotenoids, vitamin C and polysaccharides.

Polysaccharides are polymeric carbohydrate molecules composed of long chains of monosaccharides units bounds together by glucosidic bonds that generate various structures that determine different biological properties such as antioxidant, immunomodulatory, antitumor, antiinflammatory, antidiabetic, and antifatigue effects for which they have attracted growing interest to be studied. Hence, this chapter aims to describe the most relevant information on bioactive polysaccharides reported in tropical fruits and their relationship with potential beneficial health effects.

Chapter 9 - Marine natural product discovery has been acknowledged as the most important multidisciplinary endeavor of bioactive substances and drug leads. Marine flora and fauna are taxonomically different with huge productivity and are pharmacologically active novel chemical signatures that account for more than 90% of the total oceanic biomass. Marine organisms continue to amaze us for their capacity to produce chemically unique and biologically important small molecules. Marine products mediate key interactions which has improved human health as medicines and has helped to illuminate many foundational cellular processes. The major effort in marine product discovery is focused on compounds of cosmetics and pharmaceutical interest. Several promising compounds derived from marine sources showed great potential for development as pharmaceuticals. In the recent past, numerous marine natural therapeutic compounds have been isolated, characterized, identified and are under trials for human uses. The marine drugs have a combination of unprecedented chemical structures, biological activities of therapeutic interest for urgent medical needs and new molecular targets that provide the fuel that sustains. In this chapter, the authors have tried to collect the significance of marine organisms in natural product research and the application of emerging drugs to this field.

Chapter 10 - There is an increasing demand for eggs and chicken meat around the world, due to their content in high quality nutrients, such as vitamins, minerals and proteins at a very low price compared to the other animal products. Nowadays, a continuous raise of production cost is evident even in poultry due to the increase in the prices of soybean and cereal grains. As a result, non-conventional low input feeding strategies based on agro-industrial by-products (AIBP) appear as a prerequisite for ensuring poultry efficiency. The application of several AIBP in poultry diets could combine the improvements in growth performance and quality characteristics of eggs and meat with the prevention of hazards that are associated with their disposal into the environment. At the same time, enrichment of eggs and meat with bioactive compounds that are contained in AIBP fortifies human organism against the adverse effects of free radicals and oxidation, while production cost is reduced and farmer profit is increased. AIBP potential utilization in commercial poultry production requires their proper management, identification of their nutritive value and application of appropriate processing methods. Nevertheless, elucidation of the metabolism of AIBP bioactive compounds in the organism of hens and broilers is essential and will give us the opportunity to adjust the optimal AIBP supplementation levels in hens and broilers and establish their regular use.

Chapter 11 - As a response to external stress, to prevent damage caused by reaction oxygen species (ROS) in the plant cells, they generate antioxidant molecules able to help them fight those adverse conditions. Lycopene is a potent antioxidant found mostly in red fruits and vegetables such as tomato, watermelon or grapefruit. The intake of this bioactive compound has been associated with many health benefits, and therefore the objective of this chapter is to discuss the role of lycopene for an improved health, while also highlighting the most valuable natural sources of this molecule.

The results of the review showed that tomato and tomato products are among the richest sources of lycopene, and that some processing operations can in fact enhance the lycopene content and bioavailability, by favoring the transformation of trans into cis forms of lycopene, better absorbed by the human body. Regarding the health effects, it was found scientific evidence of the beneficial action of lycopene on a variety of diseases, including diabetes, obesity, cardiovascular diseases, vascular health, neurodegenerative diseases and cancer.

In: Bioactive Compounds …
Editors: G. Yildiz and G. İzli
ISBN: 978-1-68507-361-9

Chapter 1

THE EFFECTS OF DRYING PROCESS ON BIOACTIVE COMPOUNDS OF FRUITS AND VEGETABLES

***Gulcin Yildiz*[1,*] *and Gökçen İzli*[2]**
[1]Department of Food Engineering, Igdir University,
Iğdır, Turkey
[2]Department of Food Engineering, Bursa Technical University,
Bursa, Turkey

ABSTRACT

Many bioactive compounds consist of antioxidants, vitamins, carotenoids, flavonoids, and other phenolics that exist in fruits and vegetables in appreciable amounts. Bioactive compounds have different biological activities such as scavenging reactive oxygen species (ROS), inhibiting oxidation and microbial growth, and reducing the risk of chronic diseases such as cancers and cardiovascular disorders.

[*] Corresponding Author's Email: gulcn86@gmail.com.

However, they have low stability against environmental conditions. Therefore, the drying process emerged to decrease their sensitivity and present a target delivery system. Dried fruits and vegetables can be easily stored and transported at a relatively low cost. They have reduced packing costs, and their low moisture content delays microbial spoilage. This review provides an overview of recent findings on the effects of different drying techniques on major bioactive compounds of fruits and vegetables. In particular, changes in ascorbic acids, carotenoids, flavonoids, phenolic acids, total phenolics, and antioxidant capacities are discussed in detail.

Keywords: bioactive compounds, drying process, antioxidant capacity, total phenolics, ascorbic acid

INTRODUCTION

Throughout history, people have developed many methods to keep food for a long time. The physical, chemical, biological, or microbiological destruction of foods needs to be controlled. Fresh fruits and vegetables with a moisture content of more than 80% are classified as perishable foods [1]. It is aimed at preventing the development of microorganisms and ensuring their durability by reducing the rate of 80-95% of the water in fruits and vegetables to 10%-20%. The cost of packaging, storage, and transportation is reduced due to the reduction of the moisture content as well as the reduction of volume and weight. Since the activity of microorganisms decreases, the shelf life of the product increases, and food is consumed in four seasons.

The drying of fruits and vegetables affects the food quality in terms of physical, chemical, and nutritional values. There are many drying techniques, such as sun drying (SD), freeze-drying (FD), convective drying (CVD), microwave drying (MD), heat pump (HP) drying, infrared drying (ID), radio frequency (RF) drying, refractive window (RW) drying, vacuum microwave drying (VMD), vacuum microwave drying with convective drying (VMD-CD), and fluid bed drying (FBD).

Freeze drying (FD), known as lyophilization, is an advantageous technique for producing high-quality food powders, fruits, and vegetables. Since it operates at low temperatures and high vacuum, it is a preferred method for drying foods that contain thermally sensitive and oxidation-prone compounds. Freeze drying, despite the long drying time and high cost, is the method in which the food quality is at the maximum level and the bioactive component loss is the least compared to other drying techniques. FD is considered as a standard or reference drying method in most studies [2]. Freeze drying is used in various plant-based food materials such as apple, guava, strawberry, blackberry, pumpkin, tomato, asparagus, coffee, tea, garlic, and ginger. As a result of the research, significant vitamin C losses are observed in tropical fruits as a result of drying (3-70%) [3]. Asami et al. [4] stated that the freeze-drying method has higher total phenolic compounds in strawberries and corn compared to the hot air drying method. In another study, it has been noted that blueberries and raspberries that have been freeze-dried have higher anthocyanin, phenolic compounds, and antioxidant capacities, and even increases in the number of bioactive compounds in some cases, compared to their counterparts that have been treated with air drying [5]. As a result, the freeze-drying method was recommended to retain vitamin C and phenol content. Recently, the microwave-assisted freeze-drying technique has been emphasized, and a similar or better quality product is obtained compared to the traditional freeze-drying method. On the other hand, in atmospheric freeze-drying (AFD) applied with inert gases at atmospheric pressure, the drying time is very long. It is combined with other techniques such as fluid bed and spray freeze dryers to shorten drying time. More recently, power ultrasound has proven to be an effective, environmentally friendly, and non-toxic way of accelerating the AFD process.

Convective drying (CD) is one of the most common methods of drying fruits and vegetables. The hot air oxidation applied in this method affects the content and amount of bioactive components. Convective drying causes degradation in the taste, color, and nutritional values of fruits and vegetables. In general, hot air drying significantly reduces the quality of products in terms of total polyphenol content and antioxidant capacity compared to sun drying processes [6]. The disadvantage of convective drying is the long processing time and the required energy.

Overall, the retention of flavor substances and other bioactive components sensitive to heat and oxygen, such as vitamins A, C, and thiamine, which greatly affect the organoleptic quality of dried products, is very effective in determining the methods used in drying fruits and vegetables [1]. Appropriate pre-treatments, determination of proper drying methods, and optimization of drying conditions should be determined to decrease the nutrient losses during drying.

The Effect of Drying Technology on Bioactive Compounds

The drying process has important effects on the bioactive compounds of fruits and vegetables. Below are some studies which show these effects:

In the work of Tulasidas et al. [7], grape samples were treated with and without chemical (alkaline solution) and compared the drying characteristics of treated and untreated dried grapefruits. Dried grape samples with alkaline solution resulted in a shorter drying time and better-dried product quality, including higher bioactive compounds. Microwave-dried grape samples showed higher bioactive content compared to the convective alone treatment. On the other hand, the highest bioactive content was achieved for the combined microwave-convective drying method.

Shi et al. [8] have conducted research to determine the effects of several drying methods on tomatoes' quality. Tomato samples were dried at a rate of 3-4% moisture content. The number of all-trans and cis isomers was determined. The tomato samples were dried by hot air drying at 95°C for 6-10 hours, by vacuum drying at 55°C for 4-8 hours, and by osmotic drying in sucrose solution at 65°C for 4 hours at a temperature of 25°C. During the drying process, there were no significant changes determined in the amount of lycopene (all-trans). In addition, the highest lycopene loss and formation of cis isomers were detected in tomato samples dried by hot air drying. On the other hand, combined osmotic and vacuum drying methods compared to vacuum and hot air alone drying methods resulted in less lycopene and cis-isomer formation. The main reason for this might be preventing the contact of oxygen with lycopene by sugar solution during osmotic drying and hence prevent lycopene oxidation.

Zanoni et al. [9, 10] studied the heat loss because of the oxidation in Rita variety tomatoes dried at 80°C to 110°C air temperatures and 1.5 m/s airspeed conditions. In the study, the tomatoes were divided into 2 pieces, the seeds and parenchyma sections were removed, and then the tomatoes were lined up in perforated trays and placed in a cabinet-shaped dryer. During drying, tomatoes were dried until the 10% moisture content. It has been determined that 4 hours at 110°C and 7 hours at 80°C are needed to achieve 10% moisture content for tomato samples. By removing tomato samples from trays, total dry matter, lycopene, ascorbic acid (Vitamin C) and 5-Hydroxymethyl-2-Furfural (HMF) ratios have been analyzed to calculate the heat loss depends on oxidation during the drying process. It was figured out that the ascorbic acid content was affected significantly by drying temperature. While there was no ascorbic acid determined in the tomato samples dried at 110°C with 50% moisture content, tomato samples dried at 80°C and 10% moisture content showed around 10% vitamin C. In addition, it was observed that the HMF formation during drying at 110°C resulted in two times more than compared to the drying at 80°C. In terms of lycopene content, there was no significant decrease observed in tomato samples dried at 80°C. On the other hand, a significant decrease in the samples dried at 110°C was observed.

Lavelli et al. [11] observed a 2% decrease in the lycopene content of the tomato samples by drying them at 80°C. The same researchers also recorded a 0.7% decrease in the lycopene ratio of tomato pulp dried at 70°C. The decrease in the amount of lycopene in tomatoes dried at 70°C was found to be less compared to the samples dried at a temperature of 80°C.

Böhm et al. [12] dried three different strawberry varieties by microwave-vacuum (MV), convective drying (CD), and freeze-drying (FD) in their study and investigated the effects of vitamin C, anthocyanin, phenolic substances, and antioxidant capacities of strawberry samples. With the application of convective and microwave drying methods, 40% in vitamin C, 35% in phenolic substances, and 60% in antioxidant capacity compared to the initial rate was achieved. There was no significant decrease observed in strawberry samples dried by freeze-drying.

Koca et al. [13] studied the dried carrot samples with and without boiling pre-treatment at 27, 37, 47, and 57°C. Carotenoid degradation and discoloration were investigated. A significant correlation has been detected between carotenoid degradation and discoloration. The decrease in color values with increasing time and the degree of storage condition was observed. Moreover, less carotenoid degradation in carrot samples subjected to the boiling pre-treatment was detected during storage.

Özkan et al. [14] performed a study related to the spinach samples which have a 9% moisture content. They've dried these spinach samples until their moisture content is 0.1% by the microwave drying method. Eight different power levels in the range of 90 W to 1000 W have been applied for 290-4005 seconds depending on the microwave power level by microwave in the drying process. It was determined that the most suitable drying condition was achieved at 750 W of microwave power in terms of energy consumption, vitamin C, and color quality.

Kuljarachanan et al. [15] studied the changes in ascorbic acid, phenolic compounds, limonin, and antioxidant activities of lime as a result of boiling and drying applications in their study. There was a decrease in nutritional elements and antioxidant capacity of the samples after the boiling process. In the drying application, there was a rise in the amount of limonin at the beginning of the drying process, then a decrease in its later stages. A decrease in total phenolic compounds and antioxidant capacity after drying application has been determined.

Wega-Galvez et al. [16] studied the physicochemical characteristics, rehydration rate, color parameters, structural properties, ascorbic acid content, antioxidant activity, and total phenolic content of red pepper dried at the temperature of 50 and 90°C. It has been found that the increase in the drying temperature greatly affects the ascorbic acid content. The greatest ascorbic acid loss (around 98%) in dried red pepper samples was observed at a temperature of 90°C.

Madrau et al. [17] examined the effect of the drying process at 55 and 75°C on the polyphenolic substances and antioxidant capacity of different types of apricot samples. More reduction was observed in the chlorogenic acid content of the Cafona type of apricot at higher drying temperatures. The total amount of phenolic substance exhibited a significant reduction in apricot samples dried at lower temperatures. The antioxidant content of fresh Cafona apricot samples was determined to be 4 times more than the samples dried at 75°C.

In a study by Ayan [18], the properties of dried tomato samples at different temperatures, including 40°C, 50c, and 60°C, were analyzed. The lycopene amounts in the tomato samples after sun and oven drying processes were higher compared to the fresh tomatoes. But this difference has not been found statistically significant. As a result, it was concluded that the different drying methods and temperatures did not cause a significant change in the lycopene content of tomato samples.

Er [19] dried red beet samples at 50°C, 60°C, 70°C, and 80°C with two different pre-treatments: boiling in water and microwave. The effects of these drying methods on total phenolic compounds, betaxanthin, betacyanin contents, antioxidant capacity, mineral matter, and color values of red beet samples were investigated. As a result of the drying process, it was found that all drying conditions led to a significant decrease in the chemical substances of red beet samples. This decrease was found to be lower in the samples treated with pre-treatments. The most effective result in terms of methanol extraction was found to be boiling in water when comparing the effects of microwave and boiling applications on DPPH activity. In addition, in terms of temperature, the most suitable value was obtained with a temperature of 70°C. In terms of mineral matter, the reduction in calcium, iron, potassium, magnesium, manganese, sodium, and phosphorus and an increase in boron, chromium, copper, nickel, and zinc values were detected by boiling pre-treatment. In addition, an increase in the amount of all mineral substances except sodium, nickel, and phosphorus elements has been detected in the microwave pre-treated red beet samples. It wouldn't be wrong to say that the dried red beet samples at 70°C showed the best result when the effects of drying temperatures on mineral substances were evaluated. It is possible to say when all these findings are evaluated, the red beet samples with boiling pre-treatment at 70°C provided the best drying conditions.

Mrad et al. [20] conducted a study related to the changes in quality and structural properties of dried pear fruit under 5 different temperature conditions, changing from 30 to 70°C. According to the findings obtained from the research, it was concluded that drying temperatures and times have a significant effect on the color, total phenolic compounds, and ascorbic acid content of dried pear samples. Higher temperatures caused the greatest color change. Similarly, ascorbic acid degradation was more clear at high-temperature drying.

In a study conducted by Alibaş [21], grape leaves with a moisture content of 75% were dried until the humidity level reached 9% by using 13, 15, and 17 W/g microwave powers. The drying process was completed in 210 to 270 seconds. The color values and vitamin C content of microwave dried grape leaves were compared with those of fresh grape leaves. The closest color values and Vitamin C content to the fresh sample were found for the grape leaves dried by 15 W/g microwave power.

Sangwan et al. [22] examined the changes in the nutritional properties of ginger powders obtained by shade, sun, oven, and microwave drying methods. The moisture content in ginger powders was found around 3.55% in sun-dried samples and 3.78% in shade-dried samples. The highest beta carotene and Vitamin C content were determined in the ginger powder samples dried in the shade. The polyphenol content of ginger powder samples dried by shade, sun, oven, and microwave methods was found to be 12.5, 11.8, 12.4, and 12.4

mg/100 g, respectively, and exhibited similar values. On the other hand, ginger powders subjected to different drying methods exhibited good results in terms of sensory and nutritional quality.

Wojdyło et al. [23] examined the effect of convective and microwave-vacuum drying methods on some quality parameters of cherry fruits. These parameters are determined as phenolic compounds, antioxidant capacity, and color. They used convective drying at temperatures of 50, 60, and 70°C, as well as microwave-vacuum drying at 240, 360, and 480 W microwave powers and vacuum pressures of 4-6 kPa. According to the results, it was concluded that the increase in temperature or microwave level caused lower phenolic compounds, antioxidant activity, and color.

In a study by Alibaş and Köksal [24], mallow samples with an initial moisture content of 5 g based on the dry matter were dried by 3 different drying methods, including microwave drying, conventional drying, and vacuum drying. Mallow samples were subjected to the drying process until they reached a moisture content of 0.10 g. Drying applications were achieved at 6.67, 8.67, 10, 11.33 W microwave powers, conventional drying temperatures of 50, 75, 100, and 125°C, and 3.7 kPa vacuum conditions at 2 different temperatures (50°C and 75°C). The best microwave power level in terms of color and Vitamin C content values was determined as 10 W/g in 6 minutes.

In a study conducted by Demiray [25], carrot and red pepper samples were dried industrially and conventionally by hot air, vacuum, and microwave drying methods in addition to microwave-assisted hot air drying and microwave-assisted vacuum drying combinations. Hot air drying was applied at 3 different temperatures, including 45, 55, and 65°C, and 20% moisture content. Carrot and red pepper samples were dried by the vacuum drying process under 21.5 and 48 kPa absolute pressures at 45, 55, and 65°C. The microwave drying was applied at 780 W microwave power. The drying properties of carrot and red pepper samples were examined. In order to determine the changes in some quality characteristics and nutritional values of the samples during the drying process, different analyses such as color, total dry matter, vitamin

C, beta carotene, and water activity were carried out. The highest loss of beta carotene was detected in the microwave-dried carrot samples while the least loss was detected in dried carrot samples under an absolute pressure of 21.5 kPa at 45°C. Vitamin C was not detected in the red pepper samples dried by microwave drying, hot air drying at 65°C, vacuum drying at 65°C, microwave-assisted hot air, and vacuum drying processes at 65°C.

To determine the properties and quality of dried date fruit samples, İzli [26] used three different methods, including convective (60, 70, and 80°C), microwave (120 W), and freeze-drying. All color parameters showed a difference depending on the drying method. The closest values to the fresh samples were determined by freeze-drying. All dried samples regardless of the drying methods showed an increase in the phenolic content and antioxidant capacity compared to the fresh date samples. Especially, the highest antioxidant capacity and total phenolic compounds were determined in microwave dried date samples. This study showed that microwave drying with the advantage of lower drying times compared to convective and freeze-drying produced high-quality dried date slices.

İzli [27] studied the effects of combined microwave-convective drying methods on pear samples. Six different drying combinations were applied as 90W-55°C, 90W-65°C, 90W-75°C, 160W-55°C, 160W-65°C, and 160W-75°C. Increasing microwave power and drying temperature caused the lower L* (lightness) values of the pear samples. The closest color parameters to the fresh pear samples were determined in the dried pear samples at 160W-55°C in the form of a ring. While the highest total phenolic content was observed in ring-shaped pear samples dried at 160W-65°C, the highest antioxidant capacity was determined in the fresh sample. It was determined that drying parameters and cutting style affect the product quality directly. Overall, the best results were obtained for the ring-shaped dried pears in terms of color and total phenolic content.

İzli et al. [28] investigated the drying kinetics of kumquat slices at four different drying temperatures (50, 60, 70, and 80°C). In addition, fresh kumquat samples were compared with the kumquat samples dried at various temperatures in terms of quality characteristics such as color, total phenolic content (TPC), and antioxidant capacity (ATC). It was stated that the different drying temperatures caused a significant change in all color parameters (L*: lightness, a*: greenness (-)/redness (+), and b*: blueness (-)/yellowness (+)). Among the dried kumquat samples, the closest color values to the fresh samples were obtained for the samples dried at 50°C. A significant decrease in the quality characteristics of dried kumquat samples was observed compared to the fresh kumquat samples. It was determined that kumquat samples dried at 70°C among the different temperatures (50, 60, 70, and 80°C) showed the highest total phenolic content and antioxidant capacity.

Horuz et al. [29] investigated the effects of microwave-convective and convective alone drying methods on the antioxidant capacity, vitamin C, color, rehydration properties, texture profile, and sensory characteristics of apricot fruit samples. Drying applications were applied at the air temperatures of 50, 60, and 70°C with 120, 150, and 180 W microwave power levels. It was determined that the microwave-convective dried apricot samples provided better results compared to the convective-dried apricot samples in terms of total phenolic content, antioxidant capacity, and vitamin C. According to the color results, apricot samples showed lower L* and b* values with increased hot air temperatures and microwave power levels. As a result of their data, they reported that microwave-convective drying might be used as an alternative method for apricot drying.

In a study by Yildiz and Izli [30], the effects of 2 different microwave power levels (90 and 160 W) and 6 different microwave-convective combinations (90 W- 55°C, 90 W- 65°C, 90 W- 75°C, 160 W- 55°C, 160 W- 65°C and 160 W- 75°C) on the color values (L*, a*, b*, C, α° and Δe), total phenolic content (TPC), and antioxidant capacity (ATC) of pomelo samples were investigated. It was observed that the drying temperatures and/or microwave power levels caused a significant change in the total phenolic content and antioxidant capacity of pomelo samples. Combining microwave-convective drying rather than microwave-alone drying resulted in shorter drying times and higher product quality.

In a study by Yildiz and Izli [2], the effect of ultrasound (US) on the physical and chemical/nutritional attributes of freeze-dried quince samples was determined. The quince samples were treated in an ultrasonic bath at three different times (10, 20, and 30 minutes). Untreated and ultrasonically pretreated quince samples were dried in a freeze dryer. Freeze-dried samples were analyzed in terms of physical properties such as shrinkage, rehydration, color, water activity, and hardness, and bioactive compounds including total phenols, ascorbic acid, and antioxidant capacity. A significant difference was observed among untreated and ultrasonically pretreated quince samples. The freeze-dried quince samples treated with a US (especially 20 min-treatment) showed higher preservation of bioactive compounds in addition to improved physical properties compared to the untreated quince samples.

İzli and Yildiz [31] investigated the effect of high-intensity ultrasound (HUS) pre-treatment on some quality characteristics of CVD quince fruit samples at various temperatures (50, 60, 70, and 80°C). CVD quince samples at 70°C with a 15-minute HUS pre-treatment showed the lowest water activity (0.36 ± 0.02). On the other hand, the quince fruit samples exposed to 15 min HUS pre-treatment shrank significantly, especially when combined with convective drying at 70°C. The dried quince samples with no HUS pre-treatment exhibited the most condensed and denser porous texture. While dried quince samples at 50°C with no HUS pre-treatment had the lowest antioxidant capacity (2.58 mol TE/g dry–weight), dried quince samples with HUS pre-treatment for 15 minutes at 70°C had the highest ATC (7.08 mol TE/g). Among the HUS-treated quince samples, the highest vitamin C content was found for the CVD quince samples at 70°C with 15 min HUS pre-treatment (43.79 mg/100 g on the dry weight.).

Yildiz [32] investigated the impact of high-intensity ultrasound (HIU) pre-treatment on the quality of microwave-dried pears. The functional properties of microwave-dried pears, including color and bioactive compounds (total phenolic content and antioxidant capacity), were compared to fresh pear samples at two power levels (100 and 200 W) with and without HIU. It was concluded that the TPC and ATC of microwave-dried pears treated with and without HIU pre-treatment under both power levels were decreased in comparison with fresh pears. The highest phenolic content was determined in an HIU pre-treated (20 min) and microwave-dried pear slices at 200 W. The ATC of microwave-dried pear samples at 200 W with HIU pre-treatment were found higher than those of the samples dried at 100 W. All color parameters (L*, a*, b*) showed a significant change depending on the drying powers and application of HIU pre-treatment. The closest color values to the fresh pear samples were found for the microwave-dried samples at 200 W with HIU pre-treatment for 20 min.

It is observed that the vitamin C content of apricot, nectarine, plum, and mango is not significantly affected by freeze-drying for 48-72 hours [33, 34]. At the same time, the ascorbic acid content of carrot, melon, peach, papaya, and watermelon resulted in a 3-fold increase in the 48-72 hour freeze-drying method [33, 34]. The reason for this increase is explained by the inactivation of ascorbic acid oxidase against the enzymatic oxidation of ascorbic acid [33].

Vitamin C loss is observed in most of the drying methods. The reason for this loss can be the use of ascorbic acid to protect the oxidation of polyphenols at high temperatures [35, 36]. Since longer drying times facilitate the oxidation of ascorbic acid, certain drying procedures result in a higher loss of ascorbic acid [37]. It has been reported that the loss of vitamin C and carotene is higher during the air-drying method compared to the vacuum and freeze-drying methods.

It is stated that some pre-treatments, such as sulfite addition, which is often used to reduce the rate of darkening in the drying process, may cause ascorbic acid loss [38]. The bleaching process, which is used to slow down or stop the enzyme and microbial activity, helps keep the vitamin content and color of the products [39]. Okra samples bleached with sodium metabisulfite solution and dehydrated in a vacuum oven showed higher amounts of ascorbic acid than air-dried okra samples.

Drying technologies such as tray drying, spray drying, rupture window (RW) drying, and freeze-drying (FD) were applied to keep ascorbic acid in asparagus, and it was noted that the retention of ascorbic acid was higher with RW and FD. Among the methods involving the use of heated air, microwave jet bed drying showed higher total ascorbic acid content [37].

In general, the total carotenoid content of fruits and vegetables shows a decrease after drying. The reason for the degradation of carotenoids after drying is the high sensitivity of this compound to oxidation. Direct exposure to heat and light causes oxidation by activating enzymes. In addition, drying methods using heat treatment cause higher degradation than other techniques [40]. During drying, the carotenoid in fruits and vegetables becomes vulnerable to heat due to the oxidation and destruction of conjugated double bonds in molecules. Park [41] stated that there is 63% carotenoid loss in carrots where microwave drying is applied. On the contrary, in some drying methods, the decreased oxygen during drying increases the stability of carotenoids. At the same time, the bleaching process before drying increases the stability of β-carotene and lycopene in drying.

Among the carotenoids, lutein is known to be a component that protects the retina from UV rays and associated damage. Lutein degradation was reported after drying fruits and vegetables. The lutein content of sun-dried figs showed a 5-77% decrease compared to their fresh counterparts. On the other hand, there was no change in the lutein content of freeze-dried cherry. Furthermore, the lutein content of air-dried broccoli increased by 84% [33, 42].

Among all carotenoids, lycopene is the most abundant bioactive ingredient. Processing conditions such as high temperatures, light, and oxygen cause adverse effects on lycopene. The degradation of lycopene in fruits and vegetables affects both product color and nutritional values. Freeze-dried apricots, carrots, cherries, nectarines, plums, red peppers, and tomatoes seem to contain 11-85% less lycopene than their fresh counterparts. A 50% increase in the lycopene rate of 48-hour freeze-dried peaches has been observed [33]. It was stated that the lycopene amount of tomatoes dried by osmotic vacuum drying was higher than vacuum drying [8]. The amount of lycopene loss is affected by inlet and outlet air temperatures, the feed rate, initial solid concentration, drying conditions, and compressed air flow rate [43].

Flavonoids in fruits and vegetables are affected by drying like other groups. While there is a decrease of 3-83% in the total amount of flavonoids in air-dried tomatoes at 40-80°C and sun-dried bell peppers at 30-35°C, the flavonoid content in plums dried in air at 90°C increases by 5% [44]. The total flavonoid content of the raw papaya dried at 60°C with the oven method for 24 hours increases while the flavonoid content of the ripe papaya decreases. Contrary to the increased flavonoid content of raw papayas compared to ripen papayas, the flavonoid content of freeze-dried ripe papayas increases about 2 times [45].

The total flavonoids in figs were changed after drying in the sun for 8 days at 31-34°C [46]. The increase in the total amount of flavonoids is explained by the fragmentation of the cell walls and the higher extractability of the compounds in the samples [47].

The major flavones in fruits and vegetables are the bioactive components of apigenin and luteolin. It is observed that the apigenin component is reduced by 16-33% in bell peppers and figs dried in the sun for 8-14 days at 30-35°*C*. While the luteolin-8-C-glucoside component was not found in fresh figs, it was formed in dried figs by oven and sun-drying [48].

The most widely known anthocyanidins are cyanidin, delphinidin, malvidin, pelargonidin, and peonidin. More than half of cyanidin loss was observed in plums, cashews, and red guava dried at 70-90°*C* with hot air. In addition, drying with hot air for 7 hours at 47°*C* and vacuum drying at 20-50°*C* for 24 hours caused a 45-66% reduction in the cyanidin-3-O-galactoside content in apple samples [36].

Phenolic acids are one of the bioactive compounds. They have antioxidant, antibacterial, antiviral, anticarcinogenic, anti-inflammatory, and vasodilator effects. Phenolic acids are classified as hydroxycinnamic acids and hydroxybenzoic acids. Hydroxycinnamic acids in fruits and vegetables are aromatic compounds. The most common hydroxycinnamic acids are chlorogenic, cinnamic, p-coumaric, ferulic, and sinapic acids. As a result of freeze-drying, the caffeic acid content of cabbage decreases by 28-73% while the caffeic acid content of plums dried at 90 °*C* with air increases by 11-19% [44]. It has been reported that the chlorogenic acid content was 3 to 11 times higher in figs dried in the oven [48]. While the cinnamic acid content of freeze-dried jujube was reduced by 15-69%, hot air-dried mulberry contained 8-35% more cinnamic acid [49, 50]. In addition, the p-coumaric and ferulic acid contents of hot air and freeze-dried cabbage and jujube decreased by 33-100%, while the content of p-coumaric acid and ferulic acid in sun-dried jujube increased 1.4 times [49].

CONCLUSION

Bioactive compounds have an important role in the human diet because of their functional properties. However, they are easily affected by physical, chemical, and enzymatic processes. This limits their usage as food additives. For this reason, different drying methods (sun drying, freeze-drying, microwave drying, etc.) are applied to keep the stability of bioactive compounds at a maximum level during storage. The main purpose of drying is to reduce water activity and extend the shelf life. With the drying process, water is removed from the food product and the glass transition temperature is increased. This leads to less molecular mobility and higher product durability.

This review presents the recent literature describing the effects of different drying methods on bioactive compounds of fruits and vegetables. In most cases, there is a decrease in antioxidant capacity because of oxidation and thermal degradation, while in some cases, extraordinary increases are observed. The variety of fruits and vegetables is attributed to differences in maturity, agricultural processes, external factors such as time, temperature, oxygen availability, light, and physical properties such as texture and shell thickness. In general, findings in the literature show that freeze-drying is superior to other drying methods in the protection of bioactive compounds. Compared to the negative effects of high temperatures, it causes less cell damage. In addition, non-traditional drying methods such as ultrasound, infrared drying, which are called green technologies, and their combination with other drying methods were announced as the most effective method of preserving bioactive compounds. Pre-treatments applied before drying are also an important factor in retaining bioactive compounds.

REFERENCES

[1] Sagar VR, Suresh KP. Recent advances in drying and dehydration of fruits and vegetables: a review. *J Food Sci Technol* 2010; 47 (1): 15- 26.

[2] Yildiz G, Izli G. The effect of ultrasound pretreatment on quality attributes of freeze-dried quince slices: Physical properties and bioactive compounds. *J Food Process Eng* 2019; 42(5): e13223.

[3] Marques LG, Silverira AM, Freire JT. Freeze-Drying Characteristics of Tropical Fruits. *Dry Technol* 2006; 24: 457-463.

[4] Asami DK, Hong YJ, Barrett DM, Mitchell AE. Comparison of the total phenolic and ascorbic acid content of freeze-dried and air-dried marionberry, strawberry, and corn grown using conventional, organic, and sustainable agricultural practices. *J Agric Food Chem* 2003; 51(5): 1237-41.

[5] Sablani SS, Andrews PK, Davies NM, Walters T, Saez H, *Bastarrachea* L. Effects of Air and Freeze Drying on Phytochemical Content of Conventional and Organic Berries. *Dry Technol* 2011; 29: 205-216.

[6] Melgar-Almanza A, Horacio Guzman S, Nunez-Colin C, Herrera-Hernandez MG, Medina-Ramos G. Effect of drying on antioxidant capacity, physicochemical and functional characteristics of Mexican serviceberry fruit. *J. berry Res* 2015; 5:97-105.

[7] Tulasidas, TN, Raghavan, GSV. and Norris, ER. Microwave and convective drying of grapes. *Transactions of the ASAE* 1993; 36(6), 1861-1865.

[8] Shi J, Maguer LM, Kakuda Y, Liptay A, Niekamp F. Lycopene degradation and isomerization in tomato dehydration. *Food Res Int* 1999; 32(1): 15-21.

[9] Zanoni B, Peri C, Giovanelli G, Nani R. Study of Oxidative Heat Damage During Tomato Drying. *Acta-Horticulturae* 1999a; 487: 395-399.

[10] Zanoni B, Peri C, Nani R, Lavelli V. Oxidative Heat Damage of Tomato Halves as Affected by Drying. *Food Res Int* 1999b; 31(5): 395-401.

[11] Lavelli V, Hippeli S, Dornisch K, Peri C, Elstner EF. Properties of tomato powders as additives for food fortification and stabilization. *J Agric Food Chem* 2001; 49(4): 2037-2042.

[12] Böhm V, Kühnert S, Rohm H, Scholze G. Improving the Quality of Microwave-vacuum Dried Strawberries: A Preliminary Study. *Food Sci Technol Int* 2006; 12 (1): 67-75.

[13] Koca N, Burdurlu HS, Karadeniz F. Kinetics of colour changes in dehydrated carrots. *J Food Eng* 2007; 78: 449-455.

[14] Özkan IA, Akbudak B, Akbudak N. Microwave Drying Characteristics of Spinach. *J Food Eng* 2007; 78: 577-583.

[15] Kuljarachanan T, Devahastin S, Chiewchan N. Evolution of antioxidant compounds in lime residues during drying. *Food Chem* 2009; 113: 944-949.

[16] Wega-Galvez A, Scala KD, Rodriguez K, Lemus M, Miranda M, Lopez J, Perez-Won M. Effect of air-drying temperature on physico-chemical properties, antioxidant capacity, colour and total phenolic content of red pepper. *Food Chem* 2009; 117: 647-653.

[17] Madrau MA, Piscopo A, Sanguinetti AM, Del Caro A, Poiana M, Romeo FV, Piga A. Effect of drying temperature on polyphenolic content and antioxidant activity of apricots. *European Food Res Tecnol* 2009; 228: 441-448.

[18] Ayan H. Sun dried and artificially dried tomato (*Lycopersicon esculentum*) production and changes during the process (Graduate thesis). Ankara University, Ankara, Turkey; 2010.

[19] Er T. Effects of different drying temperature on some physical and phytochemical properties of red beet (Master thesis). Selcuk University Graduate School, Konya; 2011.

[20] Mrad ND, Boudhrioua N, Kechaou N, Courtois F, Bonazzi C. Influence of air-drying temperature on kinetics, physicochemical properties, total phenolic content and ascorbic acid of pears. *Food Bioprod Process* 2012; 9: 433-441.

[21] Alibas I. Microwave Drying of Grapevine (*Vitis vinifera* L.) Leaves and Determination of Some Quality Parameters. *J Agric Sci* 2012; 18: 43-53.

[22] Sangwan A, Kawatra A, Sehgal S. Nutritional Composition of Ginger Powder Prepared Using Various Drying Methods. *J Food Sci* 2014; 51(9): 2260- 2262.

[23] Wojdyło A, Figiel A, Lech K, Nowicka P, Oszmiański J. Effect of convective and vacuum-microwave drying on the bioactive compounds, color, and antioxidant capacity of sour cherries. *Food Bioprocess Technol* 2014; 7(3): 829-841.

[24] Alibas I, Koksal N. Convective, Vacuum and Microwave Drying Kinetics of Mallow Leaves and Comparison of Color and Askorbik Acid Values of Three Drying Methods. *Food Sci Technol* 2014; 34(2): 358-364.

[25] Demiray E. Dehydration of carrot and red pepper by different drying techniques, modeling of drying characteristics and changes in some quality properties (PhD dissertation). Pamukkale University Graduate College, Denizli; 2015.

[26] İzli G. Total phenolics, antioxidant capacity, colour and drying characteristics of date fruit dried with different methods. *Food Sci Technol* 2017; 37 (1): 139–147.

[27] İzli G. Effects of different drying applications on some quality characteristics of pear fruit. *TURJAF* 2018; 6(4): 479-485.

[28] Izli G, Izli N, Taskin O, Yildiz G. Convective drying of kumquat slices: Comparison of different drying temperatures on drying kinetics, colour, total phenolic content and antioxidant capacity. *Lat Am Appl Res* 2018; 48:37-42.

[29] Horuz E, Bozkurt H, Karataş H, Maskan M. Comparison of quality, bioactive compounds, textural and sensorial properties of hybrid and convection-dried apricots. Food Measure. *J Food Meas Charact* 2018; 12: 243-256.

[30] Yildiz G, Izli G. Influence of microwave and microwave-convective drying on the drying kinetics and quality characteristics of pomelo. *J Food Process Preserv* 2019; 43(6): e13812.

[31] İzli G, Yildiz G. Evaluation of the high intensity ultrasound pre-treatment effects on the physical properties and bioactive compounds of convective dried quince samples. *Int J Fruit Sci* 2021; 21(1): 645-656.

[32] Yildiz G. The Effect of High Intensity Ultrasound Pre-treatment on the Functional Properties of Microwave-dried Pears (*Pyrus communis*). *Lat Am Appl Res* 2021; 51(2):133-137.

[33] Leong SY, Oey I. Effects of Processing on Anthocyanins, Carotenoids and Vitamin C in Summer Fruits and Vegetables. *Food Chem* 2012; 133: 1577 - 1587.

[34] Shofian NM, Hamid AA, Osman A, Saari N, Anwar F, PakDek MS, Hairuddin MR. Effect of freeze-drying on the antioxidant compounds and antioxidant activity of selected tropical fruits. *Int J Mol Sci* 2011; 12: 4678- 4692.

[35] Toor RK, Savage GP. Effect of Semi-Drying on the Antioxidant Components of Tomatoes. *Food Chem* 2006; 94: 90- 97.

[36] Joshi APK, Rupasinghe HPV, Khanizadeh S. Impact of drying processes on bioactive phenolics, vitamin c and antioxidant capacity of red-fleshed apple slices. *J Food Process Preserv* 2011; 35: 453 - 457.

[37] Nindo C, Sun T, Wang SW, Tang J, Powers JR. Evaluation of drying technologies for retention of physical quality and

antioxidants in asparagus (*Asparagus officinalis*, L.). *LWT-Food Sci Technol* 2003; 36:507–516.

[38] Garcia-artinez E, Igual M, Martín-Esparza ME, Martinez-Navarrete N. Assessment of the Bioactive Compounds, Color, and Mechanical Properties of Apricots as Affected by Drying Treatment. *Food Bioprocess Technol* 2013; 6: 3247 – 3255.

[39] Korus A. Effect of preliminary processing, method of drying and storage temperature on the level of antioxidants in kale (*Brassica oleracea* L. var. acephala) leaves. *LWT-Food Sci Technol* 2011; 44:1711- 1716.

[40] Türkyılmaz M, Özkan M, Güzel N. Loss of sulfur dioxide and changes in some chemical properties of Malatya apricots (*Prunus armeniaca* L.) during sulfuring and drying. *J Sci Food Agric* 2014; 94: 2488 - 2496.

[41] Park JS, Woo MS, Kim DH, Hyun JW, Kim WK, Lee JC, Kim HS. Antiinflammatory mechanism of isoflavone metabolites in lipopolysaccharide stimulated microglial cells. *J Pharmacol Exp Ther* 1987; 320: 1237-1245.

[42] Nora CD, Muller CDR, Bona GS, Rios ADO, Hertz PF, Jablonski A, Flors SH. Effect of processing on the stability of bioactive compounds from red guava (*Psidium cattleyanum* Sabine) and guabiju (*Myrcianthes pungens*). *J Food Composit Anal* 2014; 34: 18 - 25.

[43] Goula AM, Adamopoulos KG. Effect of maltodextrin addition during spray drying of tomato pulp in dehumidified air: ii. Powder properties. *Dry Technol* 2008; 26(6):726–737.

[44] Miletic N, Mitrovic O, Popovic B, Nedovic V, Zlatkovic B, Kandic M. Polyphenolic content and antioxidant capacity in fruits of plum (*Prunus Domestica* L.) cultivars "valjevka" and "mildora" as influenced by air drying. *J Food Qual* 2013; 36:229–237.

[45] Annegowda HV, Bhat R, Yeong KJ, Liong MT, Karim AA, Mansor SM. Influence of drying treatments on polyphenolic contents and antioxidant properties of raw and ripe papaya (*Carica papaya* L.). *Int J Food Prop* 2014; 17:283-292.

[46] Kamiloğlu S, Çapanoğlu E. Polyphenol content in figs (*Ficus carica* L.): Effect of sun-drying. *Int J Food Prop* 2014; 18(3): 521–535.

[47] Çapanoğlu E. Investigating the Antioxidant Potential of Turkish Dried Fruits. *Int J Food Prop* 2014; 17: 690 - 702.

[48] Slatnar A, Klancar U, Stampar F, Veberic R. Effect of drying of figs (*Ficus carica* L.) on the contents of sugars, organic acids, and phenolic compounds. *J Agric Food Chem* 2011; 59:11696 - 11702.

[49] Gao QH, Wu CS, Wang M, Xu BN, Du LJ. Effect of drying of jujubes (*Ziziphus jujuba* Mill.) on the contents of sugars, organic acids, α-tocopherol, β-carotene, and phenolic compounds. *J Agric Food Chem* 2012; 60:9642–9648.

[50] Rodriguez K, Ah-Hen K, Vega-Galvez A, Lopez J, Quispe-Fuentes I, Lemus M, Galvez-Ranilla L. Changes in bioactive compounds and antioxidant activity during convective drying of murta (*Ugni molinae* T.) berries. *Int J Food Sci Technol* 2014; 49: 990 – 1000.

In: Bioactive Compounds ...
Editors: G. Yildiz and G. İzli
ISBN: 978-1-68507-361-9

Chapter 2

ANTIOXIDANT CAPACITY AND TOTAL PHENOLIC CONTENT OF DRIED RED CAPIA PEPPER DURING STORAGE

***Dilek Cemal*[1] *and Gulcin Yildiz*[2,*]**
[1]Igdir University, Department of Organic Agriculture Management, Iğdır, Turkey
[2]Igdir University, Department of Food Engineering, Iğdır, Turkey

ABSTRACT

This work was undertaken to evaluate the effects of different drying temperatures (50, 60, and 70 °C) and cutting types (ring and cube cutting) on the total phenolic content and antioxidant capacity of red capia pepper during storage (14 days) both at room temperature and cold storage. Convective drying (CVD) at 50, 60, and 70 °C was applied to get dried red capia peppers. The effects of CVD on selected

* Corresponding Author's Email: gulcn86@gmail.com.

bioactive compounds (total phenolic content and antioxidant capacity) of thirteen dried red capia peppers were compared. In addition, the effects of the cutting types (ring and cube) on the final quality of pepper samples were determined both at room temperature and cold storage during two weeks (14 days). The pepper samples dried at 60 °C in ring forms (6R-CS samples) resulted in the highest antioxidant capacity and total phenolic content of all storage times. Overall, dried red capia peppers in ring forms at 60 °C can be used to acquire a high-quality food product with an enhanced total phenolic content and antioxidant capacity.

Keywords: antioxidant capacity, bioactive compounds, drying, red capia pepper, total phenolic content

INTRODUCTION

Pepper belongs to the Capsicum genus of the Solanaceae family and grows in the United States, South America, Peru, Bolivia, Costa Rica, Mexico, and almost all Southern European countries [1]. In addition to being rich in vitamins (especially A and C), minerals, phenolic components, and carotene, it also has antioxidant properties [2]. Capia pepper is a type of pepper with a long conical structure, meaty, red color, and a sweet flavor. It has been used as "paste" and "oiliness" for years [3]. Although Capia pepper is consumed fresh and dried, it is also used in the production of canned, tomato paste, pickled peppers, frozen food, and sauce. It is also used in the production of antibiotic raw materials, feed materials, and dyes [4, 5]. Like many fruits and vegetables, peppers can also be preserved for a long time by drying. Drying is one of the oldest methods known for the preservation of food. The purpose of drying is to remove free water from the product and to store the products for a long time. So, the crops can be available in areas besides the harvest season. In addition, the volume and weight of dried products are reduced, making packaging and transportation easier at a low cost. The most popular method of drying food is the convective drying method [6]. However, this method causes deterioration of taste, color, and nutritional compounds in food products [7]. In this sense, the importance of the temperature applied during the drying process is understood. This research, for this reason, has the primary aim of contributing to the work on the use of convective drying on capia pepper quality. Specifically, the effects of different drying temperatures (50, 60, and 70 °C) and cutting types (ring and cube cutting) on the total phenolic content and antioxidant capacity of red capia pepper during storage (14 days) both at room temperature and cold storage were evaluated.

Material and Methods

Sample Preparation

The red capia peppers were obtained at a local market in Iğdır, Turkey, and kept at 4 ± 0.5 °C until the experiments. They were washed to remove the dust, chemical residues, and attached dirt. After removing the excess water from the surface of pepper samples with a paper towel, the peppers were cut into two different forms (ring and cube) with a sharp knife. The beginning moisture content was defined as 83.8 ± 0.54% by drying the pepper samples at 105 ± 5 °C before reaching the stable weight in a forced air convective oven.

Drying Process

Convective drying was applied in a lab convective oven (Arçelik KMF 833I, Turkey) following the method proposed by İzli [8]. The ring and cube-shaped capia pepper samples were located in a thin layer. The air velocity was set at 1 m/s for the drying process, with air temperatures of 50, 60, and 70 °C. The drying process was continued until 12% moisture content. About 100 g of pepper samples were used for each drying experiment. The experiments were conducted with three replications. A short definition of the treatments is tabulated in Table 1.

Table 1. Treatments used in the study

Sample names	Treatments
Control	Fresh sample, no treatment
5R-CS	Peppers dried at 50 ^{O}C in ring forms (cold storage)
5R-RT	Peppers dried at 50 ^{O}C in ring forms (room temperature)
5C-CS	Peppers dried at 50 ^{O}C in cube forms (cold storage)
5C-RT	Peppers dried at 50 ^{O}C in cube forms (room temperature)
6R-CS	Peppers dried at 60 ^{O}C in ring forms (cold storage)
6R-RT	Peppers dried at 60 ^{O}C in ring forms (room temperature)
6C-CS	Peppers dried at 60 ^{O}C in cube forms (cold storage)
6C-RT	Peppers dried at 60 ^{O}C in cube forms (room temperature)
7R-CS	Peppers dried at 70 ^{O}C in ring forms (cold storage)
7R-RT	Peppers dried at 70 ^{O}C in ring forms (room temperature)
7C-CS	Peppers dried at 70 ^{O}C in cube forms (cold storage)
7C-RT	Peppers dried at 70 ^{O}C in cube forms (room temperature)

Pepper Extract Preparation

In a test tube, 500 mg of fresh and dried capia peppers were weighed and extracted 3 times with 10 ml of methanol: water (80:20, v/v) and shaken for 10 minutes in each step. Then, the tube was centrifuged at 3500 rpm for 3 minutes to get a clear extract and it was combined in another tube. The analysis of total phenolic content and antioxidant capacity was carried out by using the capia pepper extracts.

Total Phenolic Content (TPC)

The TPC of peppers in agreement with the approach by Yildiz [9] was determined with slight changes. Briefly, a part of the extract was filtered through a 0.45 µm nylon filter and 1.6 ml of Folin-Ciocalte's reagent was mixed into 0.4 ml of filtered extract accordingly diluted by methanol: water (80:20, v/v) in a test tube. The contents were mixed by vortexing and left to sit for 5 minutes. Following that, 1.6 ml of 20% Na_2CO_3 was put into the mixture and vortexed. The tubes were left at room temperature for 90 minutes. Subsequently, the mixture was centrifuged at 3500 rpm for 3 minutes (Hettich Universal 320 R, Germany) and transferred into a cuvette. The analysis was carried out by a UV-VIS spectrophotometer (Mecasys Optizen Pop, Korea) at 765 nm, and quantification was figured out by a Gallic acid calibration curve, which built a range from 0 to 30 μg/ml. The findings were demonstrated as mg GAE/100 g of dry weight.

Antioxidant Capacity (ATC)

The ATC of capia peppers was measured based on the DPPH method [10] with small changes. A stock solution of DPPH was arranged by dissolving 10 mg of DPPH in 25 ml of ethanol and diluted with 25 ml of distilled water. Then, the working solution of DPPH was arranged by 400 ml of water: ethanol (1:1, v/v) having absorbance values around 0.75-0.80 at 525 nm. 0.2 ml of pepper extract was put in a tube and mixed with 3.8 ml of DPPH solution. The tube was put into an orbital shaker. It was mixed for 26 minutes at 350 rpm at room temperature. Following centrifugation at 3500 rpm for 3 minutes (Hettich Universal 320 R, Germany), the supernatant was obtained, and the absorbance was measured at 525 nm using a Mecasys model Optizen POP variable wavelength UV-vis spectrophotometer (Korea). All measurements were performed in 30 minutes after reacting with extract and DPPH solution. The ATC was expressed as μmol of Trolox equivalent (TE) per 1 g d.w. of pepper samples.

Statistical Data

Statistical analyses were managed using a randomized plot factorial experimental design. The results were analyzed using the JMP (Version 7.0, SAS Institute Inc., Cary, NC, USA). Differences among the mean values were obtained by Fisher's least significant difference (LSD) test at $\alpha = 0.05$.

RESULTS

Total Phenolic Content (TPC)

Table 2 shows the results of drying on the total phenolic content of capia pepper samples. The total phenolic content of fresh (untreated) pepper samples was found to be 674.83 ± 1.73 mg GAE/100 g d.w for the first day of samples. The drying treatments significantly influenced the total phenolic content ($p< 0.05$). All convective-dried pepper samples showed significantly higher total phenolic content compared to the fresh sample (Table 2). An increase in the drying temperature from 60 °C to 70 °C decreased the total phenolic content significantly. As a result of an increase in drying temperatures, the reduction in the total phenolic content has also been announced in several works by researchers in pears [11] and apple samples [12]. All dried samples showed a significant reduction in total phenolic content during storage. Dried pepper samples on day 7 showed a significantly lower total phenolic content compared to the first-day samples for all treatments. Similarly, dried capia pepper samples on day 14 showed a significantly lower total phenolic content compared to the seventh-day samples for all treatments. The reduction in total phenolic content during the drying period could be related to the organization of polyphenols with other compounds (i.e., protein). Capia pepper samples dried in ring forms showed significantly higher total phenolic content compared to the dried pepper samples in cube forms. In addition, preserving the dried samples in cold storage rather than at room temperature led to significantly higher total phenolic content (Table 2). When all factors and treatments are taken into account, it is easy to conclude that the capia pepper samples dried at 60 °C in ring forms and stored in cold storage (6R-CS samples) had the highest total phenolic content of all drying times. The effects of drying methods on the phenolic substances of foods were studied previously. According to some studies, a thermal process is very suitable for increasing the phenolic content in a variety of food materials, including dry raisin [13] and apricot fruits [14]. On the other hand, while some studies [14, 15] showed that total phenolic substances decreased during heat application on food materials, some other studies pointed out that there were no notable differences [16]. So, the influence of drying methods on the phenolic substances of different food products might not produce the same and/or similar results. By taking into account the results, we can say that the drying application has changeable impacts on the phenolic substances. The physical and chemical reasons for their effectiveness in the nutritional and functional characteristics of dried food products are still being researched. The highest total phenolic content observed in the dried pepper samples at 60 °C could be as a result of more cell degradation and rupture, so this can lead to more phenolic substances being released. Moreover, this observation was probably because of exposing dried pepper samples at 60 °C to less thermal effect compared to the pepper samples dried at 70 °C.

Table 2. The effects of different drying treatments on the total phenolic content and antioxidant capacity of red capia pepper samples

Sample names	Storage (Days)	Total Phenolic Content (mg GAE/100g)	Antioxidant Capacity (μmol TE/g)
	0	$674.83 \pm 1.73^{k\,(x)}$	$4183.24 \pm 0.12^{a\,(x)}$
Control	7	$548.18 \pm 0.01^{e\,(y)}$	$3465.57 \pm 0.15^{c\,(y)}$
	14	$429.21 \pm 0.95^{\imath\,(z)}$	$2786.46 \pm 0.67^{c\,(z)}$
	0	$1123.89 \pm 0.45^{e\,(x)}$	$3580.43 \pm 0.17^{c\,(x)}$
5R-CS	7	$1014.45 \pm 0.58^{c\,(y)}$	$3048.39 \pm 0.95^{e\,(y)}$
	14	$973.76 \pm 0.14^{b\,(z)}$	$2724.19 \pm 0.76^{c\,(z)}$
	0	$989.47 \pm 0.38^{g(x)}$	$3218.17 \pm 0.24^{d\,(x)}$
5R-RT	7	$952.29 \pm 0.85^{d\,(y)}$	$3055.56 \pm 0.04^{e(y)}$
	14	$865.23 \pm 0.49^{c\,(z)}$	$2511.32 \pm 0.11^{e\,(z)}$
	0	$1078.12 \pm 0.65^{f\,(x)}$	$3276.47 \pm 0.82^{d\,(x)}$
5C-CS	7	$953.07 \pm 0.17^{d\,(y)}$	$3038.25 \pm 0.06^{e\,(y)}$
	14	$869.13 \pm 0.22^{c\,(y)}$	$2315.91 \pm 0.18^{f\,(z)}$
	0	$926.13 \pm 0.41^{h\,(x)}$	$2985.15 \pm 0.09^{e\,(x)}$
5C-RT	7	$877.45 \pm 0.56^{e\,(y)}$	$2575.76 \pm 0.68^{g\,(y)}$
	14	$798.54 \pm 0.11^{d\,(z)}$	$1985.19 \pm 0.94^{h\,(z)}$
	0	$1398.18 \pm 0.01^{a\,(x)}$	$4170.12 \pm 0.25^{a\,(x)}$
6R-CS	7	$1125.23 \pm 0.46^{a\,(y)}$	$3892.31 \pm 0.97^{a\,(y)}$
	14	$1048.15 \pm 0.62^{a\,(z)}$	$3574.16 \pm 0.65^{a\,(z)}$

Table 2. (Continued)

Sample names	Storage (Days)	Total Phenolic Content (mg GAE/100g)	Antioxidant Capacity (μmol TE/g)
6R-RT	0	1278.35 ± 0.17$^{c\,(x)}$	3780.43 ± 0.11$^{b\,(x)}$
	7	1076.87 ± 0.15$^{b\,(y)}$	3657.87 ± 0.14$^{b\,(y)}$
	14	943.74 ± 0.75$^{b\,(z)}$	3028.45 ± 0.46$^{b\,(z)}$
6C-CS	0	1315.27 ± 0.17$^{b\,(x)}$	3715.67 ± 0.14$^{b\,(x)}$
	7	1089.34 ± 0.75$^{b\,(y)}$	3170.13 ± 0.26$^{d\,(y)}$
	14	1043.52 ± 0.51$^{a\,(z)}$	2627.64 ± 0.12$^{d\,(x)}$
6C-RT	0	1235.12 ± 0.17$^{d\,(x)}$	3470.33 ± 0.41$^{c\,(x)}$
	7	955.89 ± 0.13$^{d\,(y)}$	2887.08 ± 0.49$^{f\,(y)}$
	14	858.19 ± 0.42$^{c\,(z)}$	2164.35 ± 0.01$^{g\,(z)}$
7R-CS	0	875.43 ± 0.12$^{i\,(x)}$	2716.28 ± 0.07$^{f(x)}$
	7	863.07 ± 0.67$^{e\,(y)}$	2580.08 ± 0.17$^{g\,(y)}$
	14	735.39 ± 0.15$^{e\,(z)}$	2289.23 ± 0.13$^{f\,(z)}$
7R-RT	0	796.71 ± 0.82$^{j\,(x)}$	1980.24 ± 0.73$^{h\,(x)}$
	7	685.31 ± 0.33$^{g\,(y)}$	1865.67 ± 0.32$^{i\,(y)}$
	14	611.23 ± 0.65$^{g\,(z)}$	1808.99 ± 0.91$^{i\,(z)}$
7C-CS	0	813.79 ± 0.11$^{i\,(x)}$	2038.37 ± 0.61$^{g\,(x)}$
	7	753.19 ± 0.49$^{f\,(y)}$	1958.37 ± 0.72$^{h\,(y)}$
	14	673.09 ± 0.17$^{f\,(z)}$	1746.14 ± 0.65$^{i\,(z)}$
7C-RT	0	686.09 ± 0.27$^{k\,(x)}$	1710.38 ± 0.58$^{i\,(x)}$
	7	614.18 ± 0.95$^{g\,(y)}$	1588.42 ± 0.81$^{i\,(y)}$
	14	523.25 ± 0.63$^{h\,(z)}$	953.76 ± 0.16$^{j\,(z)}$

$^{a\text{-}k}$ Treatment means showed the effects of different treatments on the same day were not significantly different ($p< 0.05$).

$^{x\text{-}z}$ Treatment means showed the effect of storage times for the same treatment was not significantly different ($p< 0.05$).

Antioxidant Capacity (ATC)

The differences in the antioxidant capacity of the pepper samples obtained by different drying applications are shown in Table 2. The antioxidant capacity in the untreated (fresh) pepper samples was determined as 4183.24 ± 0.12 μmol TE/g d.w. A significant decrease was observed in the antioxidant capacity of the pepper samples in all drying applications (Table 2). Several researchers observed a decrease in antioxidant capacity after the drying process [11, 14, 17], including strawberries [17], pepino [18], and apple samples [14]. The decomposition of antioxidant compounds during the drying process can be the reason for this reduction. While the highest antioxidant capacity was obtained for the pepper samples dried at 60 °C, the lowest antioxidant capacity was found for the pepper samples dried at 70 °C. Compared to the pepper samples dried in cube forms, the pepper samples dried in ring forms showed significantly higher antioxidant capacity (Table 2). The capia pepper samples dried at 50 °C showed significantly lower antioxidant capacity compared to the pepper samples dried at 60 °C. The results showed that drying pepper samples at lower temperatures in convective drying, which results in longer drying periods, can be the reason for the more rapid decrease in antioxidant activity. In addition, all dried samples on day 0 showed significantly higher antioxidant capacity compared to the other storage times (days 7 and 14). There might be a synergistic or antagonistic effect between the antioxidant compounds and other substances [18], for instance; lime and carrot [19], apricot [14], red pepper [20], pomelo [21], kumquat [22], quince [23, 24], pear [9]. Kanner et al., [25] showed that the higher antioxidant activity of fruit and vegetables could be correlated with the synergistic effects of natural phenolic compounds. On the contrary, the antagonistic effect between TPC and ATC was observed in the present study. The fresh pepper samples demonstrated the lowest total phenolic content and the highest antioxidant capacity (Table 2).

Final Remarks

In the present study, the convective-dried capia pepper samples at 60 °C in ring forms and kept in cold storage demonstrated higher retention of bioactive compounds, including total phenolic content and antioxidant capacity. By looking at the influences of different drying conditions and factors (drying temperature, cutting types, and storage degree) on the quality retention of the dried pepper samples, the most appropriate drying temperature, cutting type, and storage degree were determined as 60 °C, ring forms, and cold storage. The results from this study are significant for the processing of dried pepper samples with enhanced bioactive compounds to get a high-quality product by optimizing the conditions.

ACKNOWLEDGMENT

This work has been supported by the Research Fund of Iğdır University. Project Number: UBY0621Y22.

REFERENCES

[1] Kumari S. Influence of climate change in capsicum production.104-107p.*Vegetable production under changing climate scenario.* 1-21 September. Nauni /Solan; 2012.

[2] Karaağaç O, Balkaya A. Evaluation of variation and characterization of collected red pepper [*Capsicum annuum* l. var. conoides (mill.) irish] populations from bafra plain. *Anadolu J. Agric. Sci.* 2010; 25(1): 10-20.

[3] Demirel K, Genç L, Saçan M. Effects of Different Irrigation Levels on Pepper (*Capsicum annum* Cv. Kapija) Yield and Quality Parameters in Semi-Arid Conditions. *J. Tekirdag Agric. Faculty* 2012; 9(2): 7-15.

[4] Hekimoğlu B. Altındeğer M. *Capia pepper production in Samsin Samsun*. Samsun Provincial Directorate of Agriculture Strategy Development Unit; 2009.

[5] Akgün M. *Pepper Paste Production Plant Industry profile*. General directorate of industrial research and development. Turkey Ministry of Industry and Trade; 2010.

[6] Michalska A, Wojdyło A, Lech K, Łysiak G P, Figiel A. Physicochemical properties of whole fruit plum powders obtained

using different drying technologies. *Food Chem.* 2016; 207: 223-232.

[7] Calin-Sanchez A, Figiel A, Wojdyło A, Szarycz M, Carbonell-Barrachina AA. Drying of garlic slices using convective pre-drying and vacuum-microwave finishing drying: kinetics, energy consumption, and quality studies. *Food Bioprocess Tech.* 2014; 7(2): 398-408.

[8] İzli G. Effects of different drying applications on some quality characteristics of pear fruit. *TURJAF* 2018; 6(4): 479-485.

[9] Yildiz G. The Effect of High Intensity Ultrasound Pre-treatment on the Functional Properties of Microwave-dried Pears (*Pyrus communis*). *Lat. Am. Appl. Res.* 2021; 51(2):133-137.

[10] Brand-Williams W, Cuvelier M E, Berset CLWT. Use of a free radical method to evaluate antioxidant activity. *LWT-Food Sci. Technol.* 1995; 28(1): 25-30.

[11] Santos S C, Guiné R P, Barros A. Effect of drying temperatures on the phenolic composition and antioxidant activity of pears of Rocha variety (*Pyrus communis* L.). *J. Food Meas. Charact.* 2014; 8(2): 105-112.

[12] Vega-Gálvez A, Ah-Hen K, Chacana M, Vergara J, Martínez-Monzó J, García-Segovia P, Lemus-Mondaca R, Di Scala K. Effect of temperature and air velocity on drying kinetics antioxidant capacity total phenolic content colour texture and microstructure of apple (var. Granny Smith) slices. *Food Chem.* 2012; 132(1): 51-59.

[13] Carranza-Concha J, Benlloch M, Camacho M M, Martínez-Navarrete N. Effects of drying and pretreatment on the nutritional and functional quality of raisins. *Food Bioprod. Process* 2012; 90(2): 243-248.

[14] Sultana B, Anwar F, Ashraf M, Saari N. Effect of drying techniques on the total phenolic contents and antioxidant activity of selected fruits. *J. Med. Plant Res.* 2012; 6: 161–167.

[15] Zanoelo E F, Cardozo-Filho L, Cardozo-Junior E L. Superheated steamdrying of mate leaves and effect of drying conditions on the phenol content. *J. Food Process Eng.* 2006; 29(3): 253-268.

[16] Dewanto V, Wu X, Adom K K, Liu R H. Thermal processing enhances the nutritional value of tomatoes by increasing total antioxidant activity. *J. Agric. Food Chem.* 2002; 50(10): 3010-3014.

[17] Wojdylo A, Figiel A, Oszmianski J. Effect of drying methods with the application of vacuum microwaves on the bioactive compounds, color, and antioxidant activity of strawberry fruits. *J. Agric. Food Chem.* 2009; 57(4): 1337-1343.

[18] Di Scala K, Vega-Gálvez A, Uribe E, Oyanadel R, Miranda M, Vergara J, Quispe I. Changes of quality characteristics of pepino fruit (*Solanum muricatum* Ait) during convective drying. *Int. J. Food Sci. Technol.* 2011; 46(4): 746-753.

[19] Duh P D, Tu Y Y, Yen G C. Antioxidant activity of water extract of HarugJyur (*Chrysanthemum morifolium* Ramat). *Lebens Wiss. Technol.* 1999; 32: 269–277.

[20] Zhou L, Cao Z, Bi J, Yi J, Chen Q, Wu X, Zhou M. Degradation kinetics of total phenolic compounds, capsaicinoids and antioxidant activity in red pepper during hot air and infrared drying process. *Int. J. Food Sci. Technol.* 2016; 51: 842–853.

[21] Yildiz G, Izli G. Influence of microwave and microwave-convective drying on the drying kinetics and quality characteristics of pomelo. *J. Food Process Preserv.* 2019; 43(6): e13812.

[22] Izli G, Izli N, Taskin O, Yildiz G. Convective drying of kumquat slices: Comparison of different drying temperatures on drying kinetics, colour, total phenolic content and antioxidant capacity. *Lat. Am. Appl. Res.* 2018; 48:37-42.

[23] Yildiz G, Izli G.The effect of ultrasound pretreatment on quality attributes of freeze-dried quince slices: Physical properties and bioactive compounds. *J. Food Process Eng.* 2019; 42(5): e13223.

[24] İzli G, Yildiz G. Evaluation of the high intensity ultrasound pre-treatment effects on the physical properties and bioactive compounds of convective dried quince samples. *Int. J. Fruit Sci.* 2021; 21(1): 645-656.

[25] Kanner J, Frankel E, Granet R, German B, Kinsella J E. Natural antioxidants in grapes and wines. *J. Agric. Food Chem.* 1994; 42: 64–69.

Chapter Reviewed by:

Assoc. Prof. Gökçen İzli, Bursa Technical University, Faculty of Engineering and Natural Sciences, Department of Food Engineering, Bursa, Turkey

In: Bioactive Compounds …
Editors: G. Yildiz and G. İzli
ISBN: 978-1-68507-361-9

Chapter 3

ENCAPSULATION OF BIOACTIVE COMPOUNDS FROM HERBAL EXTRACTS USING SPRAY DRYING

***Marija Banožić*[1]*, Jelena Vladić*[2] *and Stela Jokić*[1,*]**
[1]University of Josip Juraj Strossmayer in Osijek,
Faculty of Food Technology Osijek, Osijek, Croatia
[2]University of Novi Sad,
Faculty of Technology, Novi Sad, Serbia

ABSTRACT

In the past few decades, the spray drying technique has gained wide acceptance as an alternative drying technique for transferring liquid forms into stable dry powder forms with application in the food, pharmaceuticals and cosmetic industry. Spray drying technique enables short operation time, drying of heat sensitive materials, desirable moisture content and desirable particle size of obtained powder. For the centuries, medicinal plants are considered as one of the most important

* Corresponding Author's Email: stela.jokic@ptfos.hr.

sources of bioactive compounds. Different plant parts including leaves, flowers, stalks, fruits, seeds and roots but also whole plants are used for producing herbal pharmaceuticals, herbal cosmetic and dietary supplements. One of the most common used form are herbal extracts, produced with different extraction techniques (conventional and green novel techniques). Herbal extracts are highly concentrated in bioactive compounds such as phenolic compounds, tannins, alkaloids, pigments, vitamins, bioactive polysaccharides, peptides and volatile compounds with different biological properties. Liquid form of herbal extracts suffers from many disadvantages, such as short shelf life, higher cost of transport and storage. This chapter provides an overview on the principal applications related to encapsulation of bioactive compounds from herbal extracts with special attention on the choice of encapsulation agent on the quality of final dry extract. The application of spray drying represents a new promising drying technique for the production of dry extracts rich in bioactive compounds, which can be easily incorporated into new highly-valuable, bioactive compounds-rich products.

Keywords: bioactive compounds, herbal extracts, medicinal plants, spray drying

INTRODUCTION

Medicinal plant and plant-derived medicines are used for the centuries across the world. The world knowledge about medicinal plants is rooted in a long history of healing treatments and health traditions dating back to ancient Chinese, Sumerian, Indian, Egyptian, Greek, Roman and Arabic medical history. And that knowledge has been over the millennia transmitted (verbal and written) from generation to generation. Some sub-regions in Europe are characterized with a high number of medicinal herbs species, especially Mediterranean region and Mountain regions such as the Alps, the Pyrenees, the Massif Central in France and the Carpathian Mountains and the Balkan and the Crimean Peninsula [1].

Today, the popularity of the plant derived medicines is gained wide acceptance as an alternative to synthetic chemicals. Such "naturally derived" products have become more widely commercially available, particularly in developed countries. Byeon et al., [2] reported that every fourth drug prescribed in the US contains at least one active ingredient of plant origin. In industry, medicinal plants are utilized for the production of extracts, phytopharmaceuticals, nutraceuticals and cosmeceuticals and their application is expected to grow [3]. Gomes [4] reported that the global herbal medicine market size was valued at USD 71.19 billion in 2016 where the extracts segment valued 27.1 billion USD. Moreover, the extracts segment is expected to reach USD 44.6 billion by 2024. Properties such as a short shelf life and an unpleasant odor limit the use of the extracts. On the other hand, a segment of tablets and capsules is expected to achieve the fastest growth among other dosage forms. Also, dry dosage form can be stored for a prolonged time, handling, transport is easier and dry form enables accurate dosing and control of organoleptic properties, especially undesirable taste.

As a result of the rising interest in the medicinal and aromatic plants, new opportunities are opening up for pharmaceutical industry, particularly in improving already existing and developing new products.

Spray drying is a widespread technique for obtaining powder products in the food industry, especially in the milk, juice and coffee industry. In the pharmaceutical industry, spray drying represents a reliable one-step technique where liquid forms are transformed into powder forms. During the process, different operating conditions can be changed, including inlet/outlet/liquid feed temperature, type and concentration of encapsulating agent, feed concentration and flow rate, air flow rate and atomization. With optimizing those parameters, it is possible to obtain the powder with well-defined properties. Diversity of plant kingdom, especially in chemical composition and use of different extraction techniques requires an appropriate selection of operating conditions of spray drying [5, 6].

This chapter focused on the most commonly used medicinal and aromatic plants. Most relevant studies dealing with the transformation of herbal extracts into dry forms using spray drying will be discussed with a special emphasis on the encapsulation of bioactive compounds, namely phenolic compounds, pigments and essential oils.

Bioactive Compounds in Herbal Extract

World Health Organisation (WHO) defines herbs as crude plant material including leaves (*folium*), flowers (*flos*), fruits (*fructos*), seeds (*semen*), stems (*culmus*), wood (*lignum*), roots (*radix*), rhizomes (*rhizome*) and other plant parts, which can be whole, fragmented, or powdered while herbal materials referred to whole plants or parts of medicinal plants in the solid state: including herbs, fresh juices, gums, fixed oils, essential oils, resins and dry powders of herbs [7].

Plants produce bioactive compounds as secondary metabolites which increase their overall ability to survive the environmental condition, or to interact with the environment [8]. In human and animal metabolism, those compounds can cause various pharmacological or toxicological effects [9]. Some of the reported beneficial biological activities of plant extracts are: antimicrobial, antioxidant, anticancer and analgesic activity [10]. Classification of bioactive compounds can be according to the synthesis pathways, pharmacological effect, while Croteau et al., [11] proposed classification in three main categories: terpenes and terpenoids, alkaloids and phenolic compounds. One of the commonly used classification is presented in Figure 1.

Bioactive compounds from plant materials can be extracted by applying conventional and innovative extraction techniques. Conventional techniques are based on the utilization of different solvents with or without heating and mixing. Some examples of the conventional extraction techniques are Soxhlet extraction, maceration and hydrodistillation. Since those methods involve consumption of toxic solvents, post-extraction process requires an evaporation/concentration step for recovery of solvent and extended time to be carried out. Also, thermal degradation of bioactive compounds due to application of the high temperatures during prolonged treatment time should not be ignored [12]. The innovative extraction techniques are based on the reduced extraction time, avoided or minimized utilization of organic solvents and lower temperatures. Some of the innovative extraction techniques are extraction with supercritical and subcritical solvents, extraction with eutectic solvents and ionic liquids, extraction assisted with ultrasound/ microwaves/electric discharges [13].

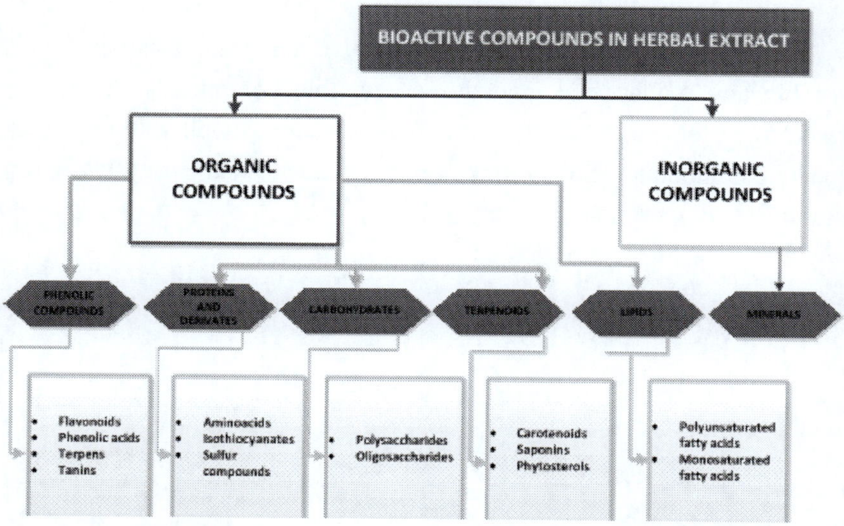

Figure 1. Classification of bioactive compounds present in herbal extracts.

Usually, the final product after the extraction is the liquid extract, which can be processed in few steps including purification of extract, fractionation, solvent evaporation and modification. The liquid form of herbal extract has many disadvantages, including a short shelf life, high cost of transport and storage. Considering the potential application of bioactive compounds in herbal extracts some efficiency issues are still questionable. Bioactive compounds can be degraded during digestion, due highly acidic pH of the stomach. Also, some bioactive compounds can be metabolized in the liver before reaching the systemic circulation [2]. Furthermore, some bioactive compounds are presented in low concentrations in plant materials and massive harvesting is needed to obtain sufficient amounts of the desired phytochemicals. On the other hand, their structural diversity and complexity often make chemical synthesis unprofitable [12]. Therefore, industry and science community are forced today to find solutions for an unstable liquid form of the herbal extract rich in bioactive compounds.

Spray Drying in Encapsulation of Bioactive Compounds

Shishir et al., [14] characterized bioactive compounds as chemically unstable and oxidative degradation-susceptible compounds. Even they are known for their health-promoting activities, application of pure bioactive compounds is very limited primary due to their fast release, low solubility and poor bioavailability.

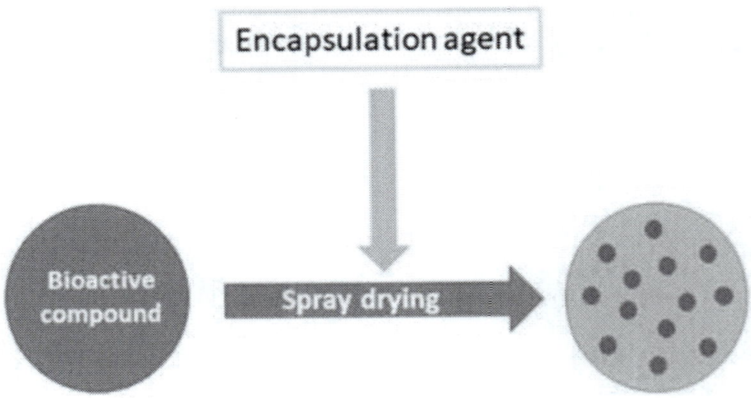

Figure 2. Form of encapsulated system.

For decades spray drying technique was considered as a dehydration process but today, encapsulation with spray drying is the most promising technology in the food and pharmaceutical industry for entrapping of bioactive compounds [15]. Encapsulation represents a process where a bioactive compound is being packed within a wall (carrier) material (Figure 2). Encapsulation products, depend on a delivery process, process condition and choice of wall material are different in size and properties. According to the particle size, obtained encapsulation product can be divided into the two groups: microcapsules (3–800 μm) and nanocapsules (< 1 μm) [16]. Obtained product is directly influenced by several process conditions, chemical and physical properties of encapsulated bioactive compound and encapsulation agent.

Reasons for the encapsulation of bioactive compounds are multiply (Figure 3). Bioactive compounds are often sensitive to elevated temperature, light and oxygen. Sometimes, the bioactive compound needs to be isolated from other compounds in the product to avoid interaction.

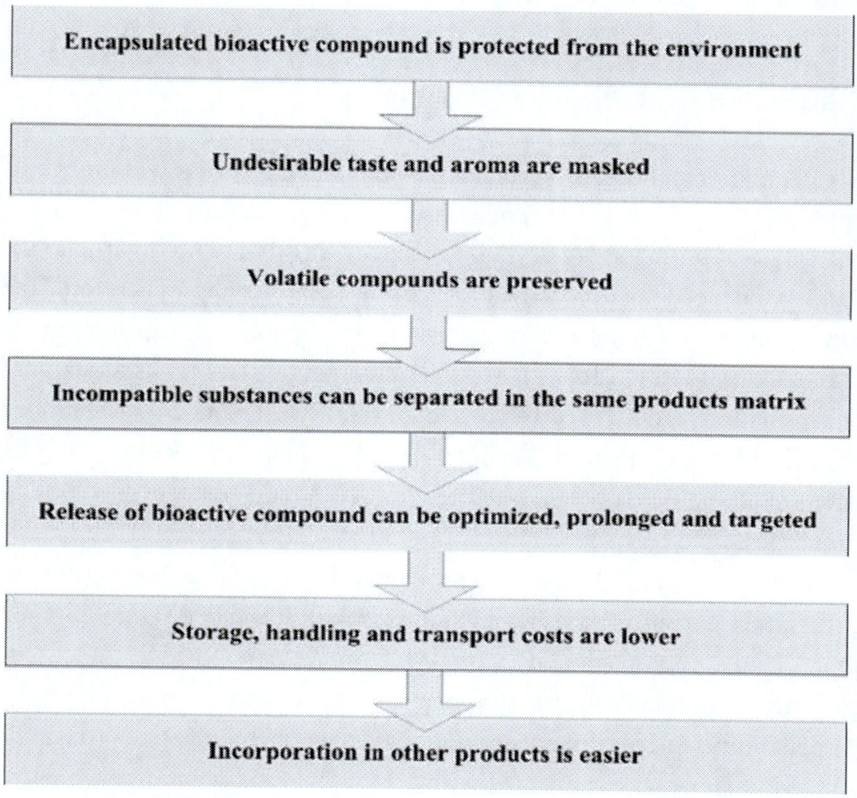

Figure 3. Summarized advantages of dry forms of herbal extracts.

Also, in pharmaceutical and cosmetic products, sometimes is necessary to ensure the controlled release of the bioactive compounds in product or organism. According to the European Directive definition [17] the controlled release is a, "modification of the rate or place at which an active substance is released." There are two main release types: delayed-release where the release of a bioactive substance is postponed and continued release where the bioactive compound has continued discharge with constant concertation [18]. Moreover, low water activity values of powder ensure microbiological, physical and chemical stability of microencapsulated compounds. Also, encapsulation helps to achieve desirable organoleptic properties, usually with masking odor and taste of bioactive compounds. In addition, dry form enables easier handling and transport, as well as lower storage and transport cost [19, 20].

Encapsulation of Phenolic Compounds

Phenolic compounds are bioactive compounds, with a wide range of polarity, from polar to non-polar, making encapsulation agent choice a critical point during spray drying. Since they mostly are not hydrophobic, there is no need for emulsification before spray drying. Therefore, the most common encapsulation agent for phenolic encapsulation is maltodextrin (MD) (Table 1). Several authors reported MD as an optimal encapsulation agent for diverse phenolics [21, 22, 23, 24]. Mentioned studies reported that utilization of MD help to achieve higher encapsulation yield, better powder properties and enhanced solubility properties.

Choi et al., [25] reported that angelica utilization is limited due to difficulties during processing including solubility and dispersibility. In their study, encapsulation with MD improved powder functional properties, particularly in the context of the water solubility and dispersibility. Successful encapsulation of phenolic compounds (decursin and decursinol) was confirmed with HPLC analysis where their content was almost two folds higher in the spray-dried powders compared to the native ones. Stability of encapsulated basil phenolics was confirmed in the study [23] where powder stability and total phenolic content was of constant quality, even after 50 days of storage.

Freeze drying is a commonly used alternative method for encapsulation of bioactive compounds. Gallo et al., [25] reported that spray drying is the preferred drying technique (among others) due to its capacity to convert a liquid into a dry powder with desired specifications (water content, particle size, solubility, bulk density and physical-chemical stability). Several studies compared spray drying and freeze-drying in encapsulation of phenolic compounds from the herbal extracts. Bilušić et al., [26] compared freeze-drying and spray drying in encapsulation of phenolic compounds from aqueous extracts from oregano, thyme, sage and rosemary. They found an increased amount of phenolic compounds in microcapsules in comparison with freeze-dried extracts. Also, they performed a two-phase digestion process (gastric and duodenal) where 80% polyphenols from spray-dried extract were preserved after the gastric phase. The release characteristics of encapsulation systems are important primarily to estimate the stability and storage periods for encapsulated compounds, but also to produce controlled release systems. Medina-Torres et al., [24] studied encapsulation of laurel infusion where feed rate made a difference in the release profiles. Samples fed at a low feed rate to the spray dryer had higher stability and slower release of encapsulated content. The slower release of bioactive compounds (especially of antioxidants) is highly desirable in the digestive tract, where an extended-release period allows the antioxidant benefits to deliver for a longer time in the body.

The release of the active ingredient is also influenced by encapsulating agent choice. Chaumun et al., [27] found that the Arabic gum is the encapsulating agent that led to a faster release of the active ingredients of laurel, while modified chitosan and sodium alginate presented a slower release.

Pigments are compounds responsible for color but also possess some health benefit properties, such as antioxidant, anticancer and antimicrobial activity. Pigments can be hydrophilic (such as anthocyanins) and hydrophobic (such as carotenoids). Different environmental factors, such as high temperature, pH changes and high-water activity can influence their degradation [28]. For hydrophilic pigments, MD is the most commonly used encapsulation agent (similar to other polyphenols). Otálora et al., [29] studied microencapsulation of betalains from cactus fruit. They obtained two betalain-rich microcapsules with MD and cladode mucilage as the encapsulating agent. Encapsulated betalains showed high stability under high-humidity and high-temperature conditions.

Table 1. Recent spray drying approaches in encapsulation of phenolic compounds and pigments

No	Common name/ Latin name	Extraction technique	Spray drying conditions	Encapsulating agent	Encapsulated bioactive compound	Main Findings	Reference
1.	Angelica/*Angelica archangelica* L.	Mixing with pH 7.4 phosphate buffered saline solution	IT: 180 °C OT: 85 °C FFR: 5 mL/min	MD	Decursin and decursinol	Improved the water solubility and dispersibility	[21]
2.	Sage /*Salvia officinalis* L.	Pressurized hot water extraction	IT: 160 °C AFR: 600 L/h FFR: 8 mL/min	MD	Phenolic compounds	Improved solubility of powder, encapsulation capacity and acceptable sensory properties	[22]
3.	Sweet basil/*Ocimum basilicum* L.	Maceration (50% EtOH)	IT: 120 °C OT: 80 °C	MD	Phenolic compounds	Physical and chemical stability during storage	[23]
4.	Laurel/*Litsea* spp.	Infusion (stirring with hot water)	IT: 140,160 and 180 °C OT: 80 °C FFR: 8-10 mL/min	MD	Phenolic compounds	Shear-thinning rheological behavior Controlled release of encapsulated compounds	[24]
5.	Chamomile/*Matricaria recutita* L.	Soxhlet extraction with EtOH	n.r.	Silicon dioxide	Apigenin-7-Glucoside	Inhibition of ligature-induced alveolar bone resorption by decreasing the inflammatory responses	[32]

No	Common name/ Latin name	Extraction technique	Spray drying conditions	Encapsulating agent	Encapsulated bioactive compound	Main Findings	Reference
6.	Gentian/*Gentiana lutea* L.	Heat-assisted extraction	IT: 90 °C OT. 50-70 °C FFR: 22 rpm	n.u.	Phenolic compounds and Gentiopicroside	Improved floating properties and *in vitro* dissolution rate of bioactive compound gentiopicroside	[33]
7.	Bay laurel/*Laurus nobilis* L.	Mixing with EtOH	IT: 115 °C OT. 65 °C FFR: 4 mL/min AFR: 40 m^3h	Modified chitosan, sodium alginate and arabic gum	Phenolic compounds	High encapsulation efficiencies Controlled release of encapsulated compounds	[27]
8.	Oregano/*Origanum vulgare* L., hyssop/ *Hyssopus officinalis* L., Sage/*Salvia officinalis* L. Lemon balm/*Melissa officinalis* L. Peppermint/ *Mentha piperita* L. Thyme/ *Thymus vulgaris* L.	Hydrodistillation and reflux extraction	IT: 130 °C OT: 35 °C FFR:13%	n.u.	Polyphenols (rosmarinic acid, caffeic acid, chlorogenic acid	Encapsulation of polyphenols Valorization of byproducts	[34]
9.	Turmeric/*Curcuma longa* L.	n.r.	IT: 150 °C OT: 87 °C FFR: 4 mL/min AFR: 40 m^3/h	pectin, MD, xanthan and inulin	Curcumin	Controlled release of curcumin	[30]

Table 1. (Continued)

No	Common name/ Latin name	Extraction technique	Spray drying conditions	Encapsulating agent	Encapsulated bioactive compound	Main Findings	Reference
10.	Marigold/*Tagetes erecta* L.	Mixing with organic solvents (n-hexane, acetone, ethanol and a mixture of EtOH and n-hexane	IT: 160-170 °C Atomization pressure: 60 hPa	MD and gum Arabic (60:40, w/w)	Carotenoids (lutein and zeaxanthin)	High encapsulation efficiency of pigments Enhanced sollubility of pigments	[31]
11.	Cactus/ *Opuntia ficus-indica* L.	Maceration with a phosphate buffer	IT: 170 °C OT: 98 °C AFR:100 m^3/h FFR: 485 mL/h	MD and cladode mucilage	Betalains (Indicaxanthin, Betanin, Isobetanin)	High encapsulation efficiency Stability during storage	[29]

IT- inlet temperature
OT-outlet temperature
FFR- feed flow rate
AFR-Air flow rate
MD-maltodextrin
n.u. not used
n.r. not reported

On the other hand, for encapsulation of hydrophobic pigments emulsification step is necessary. Emulsification can be performed with an emulsifier or with a biopolymer with emulsifying properties such as gum Arabic, modified starch, whey and gelatin [20]. Curcumin is hydrophobic polyphenol, a dominant phenolic compound in turmeric and also natural yellow pigment. Goëlo et al., [30] studied encapsulation of curcumin from turmeric with different encapsulation agents. They found the fastest release of curcumin when MD and inulin were used while xanthan provided the lowest curcumin release. Their finding suggests release profiles of curcumin, could be controlled with encapsulating agent choice. In another study [31], encapsulation of zeaxanthin and lutein from marigold extracts was performed via spray drying and freeze-drying. With spray drying uniform-size microparticles were obtained, while the encapsulated powders prepared by freeze-drying showed the broken glass shape. It can be concluded that spray drying conditions (temperature, feed flow rate, atomization rated) are influencing particle physicals properties, while encapsulation agent choice is a key factory for the encapsulation process and further application of microparticles.

Encapsulation of Essential Oils

Essential oils are a mixture of aromatic and volatile compounds obtained from plants, usually by steam distillation, hydrodistillation, but also using dry distillation, mechanical processes and supercritical fluid extraction [35]. Usually, they are highly volatile and during processing and storage losses and degradation of volatile compounds can occur. Also, essential oils are characterized with low solubility in water which limits potential applications in industry. Encapsulation can improve the physical and chemical stability of essential oil, protect them from air, light, heat and humidity influence and enhance their solubility in water [36]. Encapsulation of essential oils is usually performed in three steps:

a. Dispersion of wall material in suitable solvent,
b. Formation of core-shell dispersion or emulsion,
c. Spray drying.

Encapsulation agent choice is the most studied factor during spray drying of essential oils (Table 2). There are several demands for encapsulating agent choice: low hygroscopicity to avoid agglomeration of microparticles, emulsifying properties (due to hydrophobic properties of essential oils); solubility in commonly used solvents, desired release properties, low costs and acceptable sensory properties [37]. Even MD is most widely used encapsulation agent, it has great limitation for encapsulation of essential oils due to its low emulsifying capacity and small retention of volatiles. The study from Fernandes et al., [38] showed that gum arabic and modified starch are more efficient in retaining volatiles than MD. In the same study, the presence of inulin improved the wettability of the microcapsules, but decreased the encapsulation efficiency. However, inulin can be combined with other encapsulation agents in order to improve the solubility of particles and enable application in instant products. Encapsulation agent choice can influence the release properties of essential oil microparticles. In the study by Zhang et al., [39], the release mechanism of encapsulated lavender essential oil was described as two stages: burst release and diffusion. Authors also found that the release rate of lavender oil microcapsules could be controlled by changing the type of encapsulating agent.

Some authors revealed the combination of spray drying processes with other microencapsulation techniques where the spray dying is not used for the formation of microcapsules, but to provide stability of microencapsulated systems [37]. Complex coacervation is a type of encapsulation technique where the electrostatic attraction between two biopolymers of opposite charges (usually a protein and a polysaccharide) results with formation of coacervate. Study from Hernández-Nava et al., [36] showed that complex coacervation between gelatin and chia mucilage in combination with spray drying could be used as an alternative in encapsulation of oregano essential oil.

Table 2. Recent spray drying approaches in encapsulation of essential oils

No	Common name/ Latin name	Extraction technique	Spray drying conditions	Encapsulating agent	Encapsulated bioactive compound	Main Findings	Reference
1.	Oregano/ *Origanum vulgare* L.	Hydrodistillation	IT: 160-180 °C FFR: 5 and 7.5g/min	Gelatin and chia mucilage	Essential oil	Complex coacervation end encapsulation of essential oil	[36]
2.	Eucalyptus/ *Eucalyptus globulus* L.	n.r.	IT: 150 °C OT: 80 °C FFR: 3-4 mL/min AFR: 65-80 Nm3/h	Gum Arabic and MD	Essential oil	Suitability for pulmonary delivery systems	[40]
3.	Lavender/ *Lavandula angustifolia* L.	Hydrodistillation	IT: 150 °C OT: 110 °C AFR: 40 L/min Aspiration rate:1400 mL/min FFR: 5 mL/min	Gum acacia	Essential oil with linalyl acetate as main compound	High encapsulation efficiency Stability of encapsulated compound	[41]
4.	Lavender/ *Lavandula angustifolia* L.	Hydrodistillation	IT: 180 °C OT: 80 °C FFR: 6 mL/min	Gum acacia, sodium caseinate, gelatin, chitosan, β-cyclodextrin and polyvinyl alcohol	Essential oil	Controlled release of encapsulated essential oil	[39]
5.	Spearmint/ *Mentha spicata* L.	n.r.	IT: 110-150 °C AFR: 8 L/min FFR: 0.3 L/h	MD, gum Arabic, inulin and modified starch	Essential oil with carvone as main compound	Successful encapsulation without any chemical interaction	[42]

Table 2. (Continued)

No	Common name/ Latin name	Extraction technique	Spray drying conditions	Encapsulating agent	Encapsulated bioactive compound	Main Findings	Reference
7.	Spearmint/ *Mentha spicata* L. peppermint/ *Mentha piperita* L.	Pressurized hot water extraction technique	IT: 160 °C Aspiration rate: 100% AFR: 600 L/h FFR: 8 mL/min.	MD	Essential oil (Carvone, menthol, menthone and methyl acetate)	High encapsulation efficiency Acceptable sensory properties	[43]
8.	Cinnamon/ *Cinnamomum zeylanicum* L.	n.r.	IT: 170 °C FFR: 0.8 L/h	gum Arabic, MD and inulin	Essential oil with cinnamaldehyde as main compound	Antioxidant activity during 30 days of storage	[44]
9.	Rosemary/ *Rosmarinus officinalis* L.	n.r.	IT: 170 °C FFR: 0.9 L/h AFR: 35 L/h	Gum arabic, Modified Starch, MD, Inulin	Essential oil (1,8-cineol, camphor and –pinene)	High retention of volatile compounds	[38]
10.	Thyme/ *Thymus vulgaris* L.	n.r.	IT: 110 °C OT: 70 °C FFR: 7.5 mL/min	MD	Essential oil (Thymol, p-cymene, γ-terpinenc, carvacrol and β-caryophyllene)	Antimicrobial activity	[45]

IT- inlet temperature
OT-outlet temperature
FFR- feed flow rate
AFR-Air flow rate

n.u. not used
n.r. not reported
MD-maltodextrin

After hydrodistillation of essential oils, a hydrosol and distilled leaves as a by-product remain. Hydrodistillation by-products contain water-insoluble compounds and small amounts of essential oils [46]. Alice et al., [34] used hydrodistillation by-products of *Lamiaceae* herbs for spray drying and successful encapsulation of polyphenols. Gallo et al., [25] compared physical properties of 7 seven different medicinal plant dry extracts. They found that some properties such as particle size, morphology, powder loss on drying, hygroscopicity and flow properties as well as compact hardness were dependent on spray drying conditions while other properties such as process yield, glass transition temperature, particle and bulk densities, powder composition, compact porosity, wetting and disintegration times were dependent on herbal type.

Also, there are some limitation for using spray drying in encapsulation of essential oil regarded to uniformity of obtained particles and possible loss of low-boiling point compounds [47]. Mehran et al., [42] also reported non-uniformity of particles from *Mentha spicata*. Moreover, since spray drying of essential oils include high temperatures, from 110-180°C (Table 2), some low-boiling point aromatics can be lost during the process and affect final product quality and consumer acceptance.

APPLICATION OF ENCAPSULATED BIOACTIVE COMPOUNDS

Dry extracts can be incorporated into different pharmaceutical dosage forms such as solids (capsules, tablets, sachets), semisolids (gels, creams, pastes), or liquids (solutions and suspensions) [48]. The most commonly used pharmaceutical dosage form are tablets due to easy production (direct compression), administration convenience, accurate dosing and high stability compared to oral liquid forms [49]. Mudrić et al., [33] formulated five gastroretentive floating delivery systems with gentian dry extracts where they obtained granules with improved floating properties and prolonged-release (over 8 h) of gentiopicroside as the main bioactive compound of gentian. Tissen et al., [50] used gentian spray-dried extract for formulation mini-tablets (diameter of 1-2 mm) with excellent flow properties. Mane et al., [40] developed pulmonary delivery with spray-dried eucalyptus oil. They found these formulations promising due to the reduction in problems associated with the inhalation of eucalyptus oil. Chamomile is an herbaceous plant that is naturalized across the world and is mostly consumed as tea. Also, bioactive compounds from chamomile exhibit anti-inflammatory activities and have been recognized for their effectiveness in local therapy of different skin inflammatory manifestations [51]. Study from Guimarães et al., [32] showed that dry chamomile extract could be used in dosage forms for prevention of the inflammation and alveolar bone resorption.

Encapsulated bioactive compounds can be used also in food production. The study from Alaşalvar and Cam [43] demonstrated that microencapsulated spearmint and peppermint extracts can be used as a citric acid substitute in iced tea production. Formulated ice tea products had desirable sensory properties and enhanced total phenol content. Encapsulated essential oils can be used as an alternative to chemical preservatives, due to their antioxidant and antimicrobial activities. Encapsulated thyme essential oil showed antibacterial effect against *V. parahaemolyticus* and *V. alginolyticys*, microorganisms related to shrimp diseases [45], while in study Berktas et al., [46], peppermint dry extract improved the health-promoting effects (antioxidant and α-glucosidase inhibition) of the ice creams.

Although there are many microencapsulation techniques, spray drying has been proven to be efficient at a large scale. Also, it is important to work on better understanding of the biological activities of encapsulated bioactive compounds for their safety use in food, pharmaceutical and cosmetic products.

ACKNOWLEDGMENT

This work has been supported by Croatian Science Foundation under the project "Application of innovative techniques of the extraction of bioactive components from by-products of plant origin" (UIP-2017-05-9909).

REFERENCES

[1] IPBES. Summary for policymakers of the regional assessment report on biodiversity and ecosystem services for Europe and Central Asia of the Intergovernmental Science-Policy Platform on Biodiversity and Ecosystem Services: *Appendix 2.5* 2018, available online on 26.1.2021.(https://ipbes.net/sites/default/files/eca_ch_2_appendix_2.5_medicinal_plants.pdf).

[2] Byeon J C, Ahn J B, Jang W S, Lee S E, Choi J S, Park J S. Recent formulation approaches to oral delivery of herbal medicines. *J. Pharm. Investig*. 2019; 49:17–26.

[3] Vasisht K, Sharma N, Karan M. Current Perspective in the International Trade of Medicinal Plants Material. *Curr. Pharm. Des*. 2016; 22:999.

[4] Gomes A. *Market Analysis 2020*. 6th World congress on Medicinal Plants and Marine Drugs 2019;9.

[5] Sollohub K, Cal K. Spray Drying Technique: II. Current Applications in Pharmaceutical Technology. *J. Pharm. Sci*. 2010; 99:587–597.

[6] Manu J S, Ganesh L B, Manoj B M, Randhir C B, Shashikant B D, Chirag S B. Spray drying in pharmaceutical industry: A review. *RJPDFT* 2012; 4:74-79.

[7] World Health Organization. *WHO guidelines for assessing quality of herbal medicines with reference to contaminants and residues*. 2007.

[8] Azmir J, Zaidul I S M, Rahman M M, Sharif K M, Mohamed A, Sahena F, Omar A K M. Techniques for extraction of bioactive compounds from plant materials: A review. *J. Food Eng.* 2013; 117:426–436.

[9] Bernhoft A. *A brief review on bioactive compounds in plants.* Bioactive compounds in plants-benefits and risks for man and animals. 2010; 50:11–17.

[10] Sasidharan S, Chen Y, Saravanan D, Sundram K M, Latha L Y. Extraction, isolation and characterization of bioactive compounds from plants' extracts. *Afr. J. Tradit. Complement Altern. Med.* 2011; 8:1–10.

[11] Croteau R, Kutchan T M, Lewis N G. Natural products (secondary metabolites). *Biochemistry and molecular biology of plants*. 2000; 24:1250–1319.

[12] Gil-Chávez J G, Villa J A, Fernando Ayala-Zavala J, Basilio Heredia J, Sepulveda D, Yahia E M, González-Aguilar G A. Technologies for extraction and production of bioactive compounds to be used as nutraceuticals and food ingredients: an overview. *Compr. Rev. Food Sci. Food Saf.* 2013;12: 5–23.

[13] Cvjetko Bubalo M, Vidović S, Radojčić Redovniković I, Jokić S. Green solvents for green technologies. *J. Chem. Technol. Biotechnol.* 2015; 90:1631–1639.

[14] Shishir M R I, Xie L, Sun C, Zheng X, Chen W. Advances in micro and nano-encapsulation of bioactive compounds using biopolymer and lipid-based transporters. *Trends Food Sci. Technol.* 2018; 78: 34–60.

[15] Ezhilarasi P N, Karthik P, Chhanwal N, Anandharamakrishnan C. Nanoencapsulation Techniques for Food Bioactive Components: A Review. *Food Bioproc. Tech.* 2012; 6:628–647.

[16] Meena K S, Bairwa N K, Parashar B. Formulation and in vitro evaluation of verapamil hydrochloride loaded microcapsule using different polymer. *Asian J. Biomed. Pharm. Sci.* 2011; 1:528–538.

[17] Quality of Prolonged Release Oral Solid Dosage Forms Legislative basis Directive 75/318/EEC 1992;3AQ19a.

[18] Vinceković M, Viskić M, Jurić S, Giacometti J, Kovačević D B, Putnik P, Donsìde F, Barba F J, Jambrak A R. Innovative technologies for encapsulation of Mediterranean plants extracts. *Trends Food Sci. Technol.* 2017; 69:1–12.

[19] Ré M I. Microencapsulation by spray drying. *Dry Technol.* 1998; 16: 1195-1236.

[20] Assadpour E, Jafari S M. Advances in spray-drying encapsulation of food bioactive ingredients: From microcapsules to nanocapsules. *Annu. Rev. Food Sci. Technol.* 2019; 10:103–131.

[21] Choi K O, Kim D, Lim J D, Ko S, Hong G P, Lee S. Functional enhancement of ultrafine Angelica gigas powder by spray-drying microencapsulation. *Food Sci. Technol.* 2019; 101:161-166.

[22] Alaşalvar H, Çam M. Process for production of ready to drink iced teas from sage (*Salvia officinalis* L.) and linden (*Tilia cordata*): pressurized hot water extraction and spray drying. *Food Sci. Biotechnol.* 2018; 28:779–785.

[23] Gavarić A, Vidović S, Zeković Z, Vladić J. Influence of storage time on quality of spray-dried extracts of basil (*Ocimum basilicum* L.). *Food in Health and Disease*, scientific-professional journal of nutrition and dietetics 2019;8:6–12.

[24] Medina-Torres L, Santiago-Adame R, Calderas F, Gallegos-Infante J A, González-Laredo R F, Rocha-Guzmán N E, Manero O. Microencapsulation by spray drying of laurel infusions (*Litsea glaucescens*) with maltodextrin. *Ind. Crop Prod.* 2016; 90:1–8.

[25] Gallo L, Ramírez-Rigo M V, Piña J, Bucalá V. A comparative study of spray-dried medicinal plant aqueous extracts. Drying performance and product quality. *Chem. Eng. Res. Des.* 2015; 104:681–694.

[26] Bilušić T, Drvenica I, Kalušević A, Marijanović Z, Jerković I, Mužek M N, Bratanić A, Skroza D, Zorićm Z, Pedisić S, Nedović V, Režek Jambrak A. Influences of freeze-and spray-drying vs. encapsulation with soy and whey proteins on gastrointestinal stability and antioxidant activity of Mediterranean aromatic herbs. *Int. J. Food Sci. Technol.* 2021; 56:1582–1596.

[27] Chaumun M, Goëlo V, Ribeiro A M, Rocha F, Estevinho B N. In vitro evaluation of microparticles with *Laurus nobilis* L. extract prepared by spray-drying for application in food and pharmaceutical products. *Food Bioprod. Process* 2020; 122:124–135.

[28] Castro-Enríquez D D, Montaño-Leyva B, Del Toro-Sánchez C L, Juaréz-Onofre J E, Carvajal-Millan E, Burruel-Ibarra SE, Rodríguez-Félix F. Stabilization of betalains by encapsulation—a review. *J. Food Sci. Technol.* 2020; 57:1587–1600.

[29] Otálora M C, Carriazo J G, Iturriaga L, Nazareno M A, Osorio C. Microencapsulation of betalains obtained from cactus fruit (*Opuntia ficus-indica*) by spray drying using cactus cladode mucilage and maltodextrin as encapsulating agents. *Food chem.* 2015; 187:174–181.

[30] Goëlo V, Chaumun M, Gonçalves A, Estevinho B N, Rocha F. Polysaccharide-based delivery systems for curcumin and turmeric powder encapsulation using a spray-drying process. *Powder Technol.* 2020; 370:137–146.

[31] Kurniawan J M, Yusuf M M, Azmi S S, Salim K P, Prihastyanti M N U., Indrawati R, Brotosudarmo T H P. Effect of drying treatments on the contents of lutein and zeaxanthin in orange-and yellow-cultivars of marigold flower and its application for lutein ester encapsulation. In *IOP Conference Series: Materials Science and Engineering* 509, No. 1, 012060. IOP Publishing. 2019.

[32] Guimarães M V, Melo I M, Adriano Araújo V M, Tenazoa Wong D V, Roriz Fonteles C S, Moreira Leal L K A, Lima V. Dry extract of *Matricaria recutita* L. (chamomile) prevents ligature-induced alveolar bone resorption in rats via inhibition of tumor necrosis factor-α and interleukin-1β. *J. Periodontol.* 2016;87: 706–715.

[33] Mudrić J, Pajić M, Bigović D, Đuriš J. Development of gastroretentive floating granules with gentian root extract by hot-melt granulation. *Natural medicinal materials* 2020; 40:40–44.

[34] Alice G, Corina B, Lucia P, Sultana N, Bazdoaca C, Nicoleta D. Polyphenol content dynamics in hydrodistillation water residues of lamiaceae species. *J. Essent. Oil-Bear Plants* 2019; 22:858–864.

[35] Barbieri C, Borsotto P. Essential oils: Market and legislation. In: El-Shemy, H. Potential of essential oils. London, UK: *IntechOpen*: 2018:107–127.

[36] Hernández-Nava R, López-Malo A, Palou E, Ramírez-Corona N, Jiménez-Munguía M T. Encapsulation of oregano essential oil (*Origanum vulgare*) by complex coacervation between gelatin and chia mucilage and its properties after spray drying. *Food Hydrocoll.* 2020; 109:106077.

[37] Veiga R D S D, Da Silva-Buzanello R A, Corso M P, Canan C. Essential oils microencapsulated obtained by spray drying: a review. *J. Essent. Oil Res.* 2019; 31:457–473.

[38] Fernandes R V B, Borges S V, Botrel D A. Gum arabic/starch/ maltodextrin/inulin as wall materials on the microencapsulation of rosemary essential oil. *Carbohydr. Polym.* 2014; 101:524–532.

[39] Zhang R, Huang L, Xiong X, Qian M C, Ji H. Preparation and release mechanism of lavender oil microcapsules with different combinations of coating materials. *Flavour Fragr. J.* 2020; 35:157–166.

[40] Mane A A, Bajaj A, Khadse B, Sharma B, Jain A. Pulmonary delivery of essential oil. *J. Drug Deliv. Sci. Technol.* 2016; 5:1–17.

[41] Burhan A M, Abdel-Hamid S M, Soliman M E, Sammour O A. Optimisation of the microencapsulation of lavender oil by spray drying. *J. Microencapsul.* 2019; 36:250–266.

[42] Mehran M, Masoum S, Memarzadeh M. Microencapsulation of *Mentha spicata* essential oil by spray drying: Optimization, characterization, release kinetics of essential oil from microcapsules in food models. *Ind. Crop Prod.* 2020; 154:112694.

[43] Alaşalvar H, Çam M. Ready to drink iced teas from microencapsulated spearmint (*Mentha spicata* L.) and peppermint (*Mentha piperita* L.) extracts: physicochemical, bioactive and

sensory characterization. *J. Food Meas. Charact.* 2020; 14:1366–1375.

[44] Noghabi S M, Molaveisi M. Microencapsulation optimization of cinnamon essential oil in the matrices of gum Arabic, maltodextrin, and inulin by spray-drying using mixture design. *J. Food Process Eng.* 2020; 43: e13341.

[45] Tomazelli J O, Kuhn F, Padilha P J M, Vicente L R M, Costa S W, Boligon A A, Castellví S L. Microencapsulation of essential thyme oil by spray drying and its antimicrobial evaluation against *Vibrio alginolyticus* and *Vibrio parahaemolyticus*. *Braz. J. Biol.* 2018; 78:311–317.

[46] Berktas S, Cam M. Peppermint leaves hydrodistillation by-products: bioactive properties and incorporation into ice cream formulations. *J. Food Sci. Technol.* 2020;1–12.

[47] Carvalho I T, Estevinho B N, Santos L. Application of microencapsulated essential oils in cosmetic and personal healthcare products–a review. *Int. J. Cosmet. Sci.* 2016; 38:109–119.

[48] Lengyel M, Kállai-Szabó N, Antal V, Laki A J, Antal I. Microparticles, microspheres, and microcapsules for advanced drug delivery. *Sci. Pharm.* 2019; 87:20.

[49] Wu C Y, Seville J P A comparative study of compaction properties of binary and bilayer tablets. *Powder Technol.* 2009; 189:285–294.

[50] Tissen C, Woertz K, Breitkreutz J, Kleinebudde P. Development of mini-tablets with 1mm and 2mm diameter. *Int. J. Pharm.* 2011; 416:164–170.

[51] Patzelt-Wenczler R, Ponce-Pöschl E. Proof of efficacy of Kamillosan (R) cream in atopic eczema. *Eur. J. Med. Res.* 2000; 5:171–175.

Chapter Reviewed by:

Professor Senka Vidović, University of Novi Sad, Faculty of Technology, Novi Sad, Serbia

Krunoslav Aladić, PhD, University of Josip Juraj Strossmayer in Osijek, Faculty of Food Technology Osijek, Osijek, Croatia

In: Bioactive Compounds …
Editors: G. Yildiz and G. İzli
ISBN: 978-1-68507-361-9

Chapter 4

IMPACT OF BIOACTIVE COMPOUNDS ON ANIMALS AND THE ENVIRONMENT

Alexandros Georganas[1,*], Elisavet Giamouri[1], Christos Christodoulou[1], Alexandros Mavrommatis[1], Evangelos Zoidis[1], George Papadomichelakis[1], Panagiotis E. Simitzis[2], Eleni Tsiplakou[1], Athanasios C. Pappas[1] and Kostas Fegeros[1]

[1]Laboratory of Nutritional Physiology and Feeding, Department of Animal Science, Agricultural University of Athens, Athens, Greece
[2]Laboratory of Animal Breeding and Husbandry, Department of Animal Science, Agricultural University of Athens, Athens, Greece

ABSTRACT

The use of alternative feedstuffs in animal nutrition has been associated with several benefits regarding the environment, the quality

[*]Corresponding Author's E-mail: ageorganas@aua.gr.

of the derived animal products, and the food-feed competition. Such potential feed sources include oilseeds and their by-products seaweeds and algae, food waste and agro-industrial by-products containing a plethora of bioactive compounds, namely long-chain polyunsaturated fatty acids, essential oils, polyphenols, flavonoids, carotenoids, and other antioxidants and immunoregulating compounds. Addition to animals' diets may result in producing functional animal products that meet the global market trends and fortify both humans' and animals' health. The present chapter will focus on the role of bioactive compounds present on agro-industrial by-products, innovative feeds and biotechnologically processed derivates on animal performance, the quality of the derived animal products (eggs, meat, and milk), the livestock health status, and the associated environmental benefits.

Keywords: agro-industrial by-products, alternative feedstuffs, bioactive compounds, environmental sustainability, food waste

INTRODUCTION

Given that the world population is expected to increase from 7.7 in 2019 to 9.7 billion by the year 2050 [1], there is an urgent need of carrying out alternative strategies in order to support food security, improve the quality of food products, and mitigate the environmental burden posed by animal production through the implementation of the elements of circular economy. In line with the boost of the world population is the increase of the demand of animal products [2], which is currently overcome by the expansion of agriculture in the tropics [3, 4]. Furthermore, livestock production accounts for 14.5% of the total greenhouse gas (GHG) emissions provoked by human activities [5]. Thus, the partial substitution of the main dietary ingredients of livestock diets, i.e., soybean products and cereal grains by oilseeds and their by-products, and food waste could alleviate the burden which animal production induces on the environment. Additionally, agro-industrial by-products have detrimental effects on the environment when they are improperly disposed and therefore their valorization through animal feeding can eliminate these effects [6]. Besides that, seaweeds and algae, and oilseeds and their by-products serving as feed components can lessen methane emissions caused by ruminant production due to the bioactive constituents and other compounds they contain. Simultaneously, the food-feed competition could also be shifted in favor of the food consumption by humans, while there is great potential to improve animal products' characteristics due to the properties of bioactive compounds present in the alternative feedstuffs. Moreover, the cost of livestock diets is constantly increased due to the high prices of soybean products and cereal grains. Alternative low input feeding strategies based on the above-mentioned alternative feedstuffs therefore gain ground. The multifaceted benefits of applying feeding strategies based on alternative feedstuffs have been supported by various studies, which are described in this chapter.

FEEDING ALTERNATIVE FEEDSTUFFS TO LIVESTOCK

Oilseeds and Their By-Products: A Rich Source of Bioactive Compounds in Animal Nutrition

Oilseeds are cultivated globally while being versatile crops. Consequently, they can be identified as essential crops from an economic standpoint. Oilseeds are commonly utilized in animal nutrition, pharmaceutical industry, food sector, and cosmetics. Furthermore, they are widely known as raw materials for biodiesel production. In fact, Waseem et al. [7] accurately describe them as bio-energy crops. In animal rations, they are mostly used as high protein sources, as whole oilseeds, and meals. The major contributor to the oilseed global economy is soybean. Other oilseed crops are sunflower, Brassica, canola, rapeseed, peanut, and cotton [7] while the most utilized whole oilseeds in dairy rations are whole soybeans, cottonseeds, and sunflower seeds [8]. They have high nutritional content and constitute a rich source of bioactive compounds, linked with several benefits for both human and animal health [7]. First and foremost, several oilseeds are a rich source of n-3 fatty acids, while they contain a variety of phenolic compounds, phytosterols and phytoestrogens that are associated with potential health benefits. These bioactive compounds have both antioxidant and anti-inflammatory properties [9]. Remarkably, oilseeds represent a promising source of natural antioxidants [10]. Notable examples of oilseeds rich in antioxidant compounds are linseed, sunflower, soybean, cottonseed, and canola. More specifically, sterols are an antioxidant compound group presented in oilseeds with the ability to prevent the thermal-oxidative degradation of oils. Oilseeds also contain tocopherols and tocotrienols [11]. Besides, oil extracted meals, after processing, contain mostly phenolic acids and their derivatives, flavonoids, and lignans. The importance of these antioxidants in oils and meals derives from stabilizing lipids and oils against oxidative damage [10]. However, Tanwar and Goyal [9] accurately address that each oilseed has both phytochemical and/or antinutrient components both positively and/or negatively associated with human health. Hence, awareness should be raised when offering whole oilseeds to animal rations. In general, a small amount of oilseed production is fed to animals as whole seeds [12]. Whole oilseeds are used as energy source in animal nutrition while their meals as a protein source. The presence of toxic compounds and many of the antinutritive factors presented in the oilseeds inhibit in specific cases their inclusion in animal rations since some of them like the whole cottonseeds are not fed to non-ruminant animals. Still, another aspect that should be considered when formulating a balanced ration is the caloric density of oilseeds, since they are high in nutritional energy.

Utilizing oilseed meals in animal nutrition is considered an innovative strategy not only from a nutritional but also from an economic perspective. Oilseed meals are by-products derived from oil processing and they are acknowledged as essential protein supplements in animal nutrition. Their protein content is higher than the whole seeds, while the oil content of these meals is related to the type of process used to extract the oil. The major protein source in animal rations is soybean meal (69%) followed by rapeseed meal (13%), cottonseed meal (6%), sunflower meal (5%), fish meal (2%), and peanut meal (2%) [13]. Amongst oilseed meals, soybean has the top spot in the market share, market value, and applied nutritional value. Soybean is the most dominant oilseed crop and the most important source of protein for animal nutrition, with its protein content ranging between 30–40% [14, 15], while soybean meal is comprised of a highly digestible protein consisting of a noteworthy blend of amino acids for animal feeds [16]. Up to almost 95% of the soybean meal is being produced and utilized for livestock feed. Soybean meal apart from being the major protein source in animal diets is also a notable source of minerals and vitamins.

Up until now, due to their nutritional properties and bioactive compounds, studies evaluated the inclusion of oilseeds in animal rations and their impact on animal performance, animal products, animal health, and their environmental impact.

The Effect of Different Oilseeds and Their By-Products in Animal Performance and Product Quality

Including oilseeds in ruminant diets is a direct and effective way to alter milk [17] and cheese [18] fatty acid profile to a desirable for human health consumption direction. This is achieved by reducing the saturated fatty acids (SFA) while enriching fatty acid profile with essential unsaturated fatty acids like α-linolenic and conjugated linoleic acids (CLA) [17, 18]. More specifically, supplementing ewes' diets with flaxseed and sunflower seeds increased CLA and α-linolenic acid milk concentration [18]. In the same study, although feeding sunflower seeds did not affect milk yield, flaxseed increased milk yield. Moreover, supplementing linseed to dairy goats improved milk's fatty acid profile by decreasing SFA and increasing CLA, vaccenic acid, and α-linolenic acid [19]. Furthermore, Mitsiopoulou et al. [20] reported the positive impact of sesame seeds in goats' milk chemical composition, milk fatty acid profile and oxidative stability of milk. In dairy cows, feeding extruded linseed and linseed oil altered milk fatty acid profile and increased milk yield [21]. Also, supplementing dairy cows' diets with Camelina seeds and meal slightly decreased milk protein and milk fat content, resulted in a softer butter, and altered milk fatty acid profile to a desirable direction

[22]. Similar results were also reported when oilseed by-products like cold-pressed sunflower cake [23], rapeseed and sunflower oilcake [24], linseed cake [25], and sesame meal [26] were supplemented in ruminant diets.

In accordance with the above findings, supplementing broiler diets with oilseeds (Camelina, flaxseeds, and sunflower seeds) caused a significant increase in average body weight and average daily feed intake, and specific carcass traits [27]. In the same study, broilers that were offered flaxseed and sunflower, had improved carcass quality, a higher proportion of muscles, and low abdominal fat content. In addition, the substitution of soybean meal with rapeseed meal in pig diets did not affect average daily gain, feed intake, feed conversion ratio and carcass characteristics [28]. Moreover, supplementation with flaxseed and canola seed rich in n-3 fatty acids increased essential n-3 fatty acids in broilers' meat [29]. Besides, Cherian et al. [30] reported that supplementing layers diets with Camelina meal (15% of the whole diet) 8-fold increased the n-3 fatty acids in eggs. Similarly, Aziza et al. [31] reported that feeding camelina meal at 10%, 2.5-fold increased the n-3 fatty acids of white and dark meat, while the same level of inclusion increased thigh meats antioxidant activity. In particular, the γ-tocopherols in thigh meat were 1.5-fold increased. At the same time, this supplementation reduced thiobarbituric acid reactive substances (TBARS) in the meat during storage and cooking [32].

Hence, whole oilseeds and oilseed meals can be used in animal diets to support animal performance and at the same time due to their bioactive molecules to develop animal products that preserve their oxidative stability and promote human health.

The Impact of Oilseeds and Their By-Products in Animals' Oxidative Status and Immune System

The unsaturated fatty acids that are presented in oilseeds are prone to oxidation. However, antioxidant compounds such as those that were previously mentioned counteract this oxidation. Therefore, Tsiplakou et al. [33] recently reported that the concomitant supplementation of sesame meal with vitamin E and selenium effectively tackles goats' oxidative stress. Likewise, the partial substitution of soybean meal and maize grains with sesame seeds in goats' diets improved the goat's oxidative status [20]. Furthermore, supplementing dairy cow diets with flax meal, improves the oxidative status [34], while in dairy cows that received flax hulls, the relative expression of some genes, involved in the protection against oxidative stress damage occurring in the mammary gland and other tissues was increased [35].

Regarding the immune system, the possible positive impact of whole sesame seeds in goats' innate immunity was recently assessed in a study by [36]. Besides, dairy cows receiving sunflower seeds prepartum, positively influenced postpartum ovarian function with no effect on uterine inflammatory status or pro- and anti-inflammatory cytokine gene expression in endometrial cells [37]. Interestingly, in broiler diets, the supplementation with micronized camelina, flax and sunflower seeds, positively modified the intestinal microbial population and bacterial sensitivity to selected chemotherapeutic agents [27]. Furthermore, linseed supplementation in finishing barrows diets (10%) significantly decreased the proinflammatory cytokines gene expression [38].

Consequently, oilseeds due to their bioactive molecules can prevent possible oxidative damage and boost the immune system.

Oilseeds as a Sustainable Nutritional Strategy in Mitigating the Environmental Burden

The United States (US), Brazil and Argentina are the leading countries in the global production of soybean, accounting for approximately 80% of the global supply [16]. With the global soybean production still rising, the expanded use of land and the increasing demand for soy production demonstrates detrimental environmental issues on a global scale (deforestation). To tackle this issue, and simultaneously promote biodiversity, alternative oilseed crops with lower inputs and in less productive areas should be evaluated as alternative protein feed sources that can partially substitute soybean in animal diets.

Interestingly, since dietary fat supplementation reduces methane emissions [39], oilseeds can be used as an effective nutritional strategy in mitigating livestock's environmental burden. Beauchemin et al. [40] reported that processed oilseeds, due to their content in long-chain fatty acids are an effective strategy in reducing methane emissions. More specifically, the inclusion of oilseeds cracked or crushed reduces the daily methane emissions even though part of the oil in cracked oilseeds is rumen-protected. Examples of well-known and studied oilseeds that can reduce the methane emissions are sunflower seeds [40, 41], cottonseed [41], linseed [42], and canola seeds [40]. More specifically, supplementing linseed in goat diets in two different levels of inclusion decreased the relative abundance of the total bacteria, methanogens, and protozoa compared to goat diets that were not supplemented [42]. This inclusion significantly reduced the relative abundances of *Fibrobacter succinogenes* and *Rumunococcus flavefaciens*. In addition, the inclusion of Camelina oil in dairy cattle diets reduced the methane and carbon dioxide (CO_2) emissions [43]. Likewise, sunflower and linseed oils in fistulated sheep, reduced methane production compared to control cultures, with no significant alterations being reported in the daily gas production [44]. Furthermore, substituting soybean meal with rapeseed meal in growing-finishing pigs was reported to decrease urinary nitrogen and total nitrogen excretion [28].

Therefore, the utilization of different oilseeds and the partial substitution of soybean meal can be a sustainable nutritional strategy to mitigate livestock's environmental burden.

Seaweed and Algae: A Sustainable Feeding Strategy Rich in Bioactive Compounds

During the last decades seaweed species are used in pharmaceutical industry, cosmetology, human diets, and as animal feed. They are constituted by simple to more complex organisms, able to take advantage of sunlight to convert CO_2 into sugars and oxygen, during the photosynthesis process while they pose enormous biodiversity. This diversity is also reflected in their chemical composition observing species with great differences in protein and lipids content which are of great importance. Even though macroalgae have only a small percentage of lipids (1–5%), the majority of those are long-chain polyunsaturated fatty acids (PUFAs) which have been characterized as pivotal bioactive compounds [45].

On the other hand, microalgae are mainly constituted by autotrophs representers with the most well-known species being *Arthrospira* (formerly *Spirulina*, blue-green algae), *Chlorella, Dunaliella* and *Haematocussus*. In addition, to produce n-3 fatty acids, heterotrophic marine organisms, such as *Crypthecodinium, Schizochytrium* and *Ulkenia*, are used [46]. It is worth mentioning here, that the abyssal aquatic biodiversity makes hard the absolute classification of microalgae into phyla and orders. Remarkably, *Schizochytrium* a well-documented microalga shows structure characteristics that are also found in fungal taxa; thus, the term marine microbe is more suitable.

Macro- and microalgae have attracted serious scientific attention as animal feed due to their advantages over conventional cultivations. More specifically, microalgae, unlike conventional crops, require enriched CO_2 sources unveiling a potential of combined systems that will produce algae biomass while simultaneously they would neutralize CO_2 [47]. Furthermore, their biomass production is greater than any conventional land crop considering not only their extensive growth rate but also their minimum land demands. Last but not least, both their cultivation strategies and facilities are hard to be threatened by exogenous factors since they are not affected by environmental issues such as climate change. However, the substitution of conventional feedstuffs with algae exploiting the current technology is unaffordable from a large-scale implementation perspective. Nevertheless, their potential as feed additives rich in bioactive molecules appears to be a well-justified strategy in animal nutrition.

Amongst microalgae biomolecules, PUFA has gained a lot of interest since these vital nutrients have been correlated with human health benefits [48]. The advantage of microalgae regarding PUFA formation and accumulation lies in an efficient elongase-desaturase network that is common in many algae species [49] (Figure 1). Interestingly, microalgae appear to be a customized source of PUFA either by regulating gene expressions [50] or through cultivating manipulations [51]. From this perspective, novel biotechnological tools could orchestrate PUFAs' synthesis in microalgae showing a unique tailor-made and sustainable alternative.

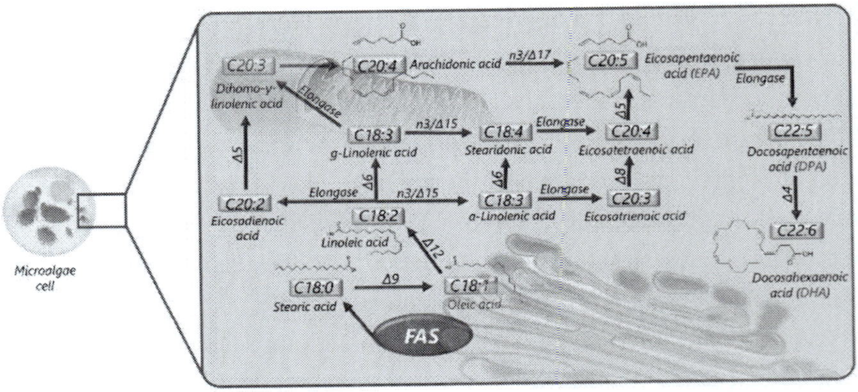

Figure 1. Biosynthetic pathways of the long-chain polyunsaturated fatty acids (LC-PUFA) in microalgae. FAS, fatty acid synthase.

Given the aforementioned advantages of seaweed and microalgae, scientists explore the impact of nutritional strategies that encompass marine organisms in livestock diets to enrich animal products with bioactive molecules. The production of functional foods with the simultaneous expansion of both livestock sustainability and animal health shapes the future livestock.

Algae in Animal Performance and Products Features

In many Western countries, the average intake of the health beneficial n-3 long chain polyunsaturated fatty acids (n-3 LCPUFA), eicosapentaenoic acid (EPA) and docosahexaenoic acid (DHA) are below the recommended level, raising interest in food enrichment with n-3 LCPUFA. In this light, in the study of [52] an EPA-rich microalga (*Nannochloropsis gaditana*) was supplemented at the doses of 5 and 10%

in hens' diets aiming to enrich eggs with n-3 LCPUFA. Microalgal EPA was barely accumulated in yolk lipids, but mainly biotransformed to DHA and being deposited in yolk phospholipids. However, the transfer efficiency of n-3 LCPUFA from *N. gaditana* to eggs was rather low [52]. Furthermore, the yolk color of n-3-enriched eggs shifted from yellow to more orange-red, probably due to the transfer of algal carotenoids to egg yolk [52]. Similarly, Petrolli et al. [53] supplemented broilers diets with 0.5, 1, and 2% DHA-rich marine microbe *Schizochytrium spp.* targeting to increase thigh meat with PUFAs. Interestingly, *Schizochytrium spp.* inclusion not only increased DHA content (0.7 vs 6.5%) in thigh meat and improved the n-6/n-3 ratio towards beneficial for human health direction, but also tended to increase broilers performance and feed conversion ratio (FCR).

Microalgae *Schizochytrium spp.* rich in DHA and docosapentaenoic (DPA) fatty acids have been also assessed as an effective nutritional alternative to enrich bovine [54], ovine [55], and caprine [56, 57] milk with bioactive lipids. However, in the study of Mavrommatis and Tsiplakou [57], the supplementation of 40 and 60 g/day *Schizochytrium spp.* in goat's diet resulted in an unfavorable reduction in milk fat content due to imbalances in rumen fermentation processes [58, 59]. Nevertheless, the lower inclusion level (20 g/day *Schizochytrium spp.*) did not adversely affect animals' performance while almost doubled milk's PUFA content (4.62 vs 8.55%) through CLA, DPA and DHA enrichment [57].

Aurantiochytrium limacinum supplementation in growing-finishing pigs resulted in a considerable upsurge in EPA and DHA content in *longissimus lumborum* muscle without affecting carcass traits and feed efficiency [60]. Additionally, *Schizochytrium spp.* dietary inclusion in finishing pigs enriched *longissimus dorsi* muscle with DHA and DPA fatty acids while improved n-6/n-3 ratio [61]. However, it could not be overlooked that an abnormal odor in the meat of pigs that were fed the *Schizochytrium spp.* was revealed while the malondialdehyde (MDA) concentration in subcutaneous backfat was considerably increased [61]. Behind the increase in MDA content, there is a well-known mechanism of PUFA oxidation. More specifically, in the study of Mavrommatis et al. [62] the dietary supplementation with 40 and 60 g/day *Schizochytrium spp.* in goat's diet compromised oxidative stability of goat milk as indicated by MDA and protein carbonyls (PC) levels. These imbalances were initiated since PUFA, and palmitic acid triggered NADPH oxidase activity leading to a cascade of prooxidant incidences [62]. Similarly, *Schizochytrium spp.* inclusion in dairy calves' diet increased serum antioxidant enzyme activities (superoxide dismutase (SOD), catalase (CAT), and glutathione peroxidase (GSH-Px)) as a response to MDA elevated levels [63]. On the other hand, supplementation with low lipids content microalgae such as *Chlorella vulgaris* [64] and *Chlorella pyrenoidosa* [65] did not burden either the goats' oxidative status or their milk oxidative stability, while in the case of *C. vulgaris* blood plasma SOD and CAT activity was increased, and milk PC levels were suppressed due to the antioxidant compounds that are present in *Chlorella* biomass.

Although the transfer efficiency of LCPUFA from algae biomass to animal products is rather low for both monogastrics and ruminants, the substitution of SFA with beneficial bio-lipids is of great importance and could contribute to a collective effort developing functional food. There are concealed however dangers compromising animal performance through metabolic imbalances such as the induction of oxidative stress which should be also addressed when farm-scale applications are designed.

Algae Protect Animal Health

Microalgae contain high levels of components, which might have biological activities and unique properties. Amaro et al. [66] summarized the potential antibacterial properties of microalgae reporting that specific fatty acids, organic acid, and extra-metabolites could effectively counteract *MRSA, E. coli, Vibrio, Salmonella, Pseudomonas,* etc. Further to their antibacterial, antiviral, and anthelmintic properties, algae biomass can be considered as a multi-component antioxidant system that in general is more effective through the interactions between the different antioxidant components. The most powerful water-soluble antioxidants in algae are polyphenols, phycobiliproteins and vitamins. In this context, El-Bahr et al. [67] observed lower values of MDA and PC in breast muscle of broilers while the activity of SOD was increased due to dietary supplementation with microalgae species (*Spirulina platensis, Chlorella vulgaris, Amphora coffeaformis*).

Contamination of animal feed with mycotoxins appears to be another common issue that threatens animals' health and product quality. Mavrommatis et al. [68] summarized the detrimental effect of mycotoxins infection on swine, poultry, and ruminants' oxidative status. Remarkably, the dietary supplementation with algae-based antioxidants minimized the detrimental effect of mycotoxin-contaminated feed and partially improved FCR in both heat-stressed and unstressed broiler chickens [69].

Without narrowing out the great importance of antioxidant compounds that are presented in algae biomass, LCPUFA such as EPA, DPA and DHA depict a pivotal role in animal health as well. More specifically, algae PUFA could regulate animals' proinflammatory response emerged by farming conditions in high-yielding individuals through the production of eicosanoids with much less proinflammatory power or by inhibiting proinflammatory cytokines signaling resulting in the suppression of low-grade inflammation and stress. Additionally, Caroprese et al. [70] showed that a mixture of phytosterols from *Dunaliella tertiolecta* reduced the cytokine production in a sheep model of inflammation. Moreover, in the in vitro study by Ciliberti et al. [71], *Chlorella sorokiniana* was found to suppress the sheep peripheral blood mononuclear cell proliferation and proinflammatory cytokine levels.

These sets of evidence demonstrate that algae and algae derivates could be useful ingredients in animal feed to minimize the dependance on antibiotics aiming to inhibit antibiotic resistance and expand animal health using natural bioactive compounds.

Algae: A Promising Ally toward Climate Change

Overlooking microalgae' superiority against conventional cultivations from a sustainability perspective, there is also a constantly increasing interest concerning their potential to mitigate methane formation in ruminants. Notably, a preliminary *in vitro* analysis suggested that red seaweed *Asparagopsis taxiformis* can reduce methane production by 95% when added to cattle diets at 5% [72]. Later, supporting the aforementioned observations, an *in vivo* study supplementing with 1% *Asparagopsis armata*, a closely related species, in dairy cows' diet showed that methane production and yield mitigated by 67 and 43%, respectively [73]. The mechanism attributed changes in methane formation lies in bromoform content in *Asparagopsis* species. This anti-methanogenic compound has been shown to inhibit enzymatic activities by binding to vitamin B12 which chemically resembles coenzyme F430 a cofactor needed for methanogenesis.

On the other hand, recent studies by Mavrommatis et al. [58, 59] have shown that DHA-rich marine microbe *Schizochytrium* spp. decreased the methanogens abundances in goat's rumen. However, in the case of lipid-rich algae, the reduction of methanogenic microbes is not attributed to the same enzymatic inhibition due to bromoform, rather to a holistic rumen microbiome shifting. These alterations not only decrease the availability of hydrogen which regulates the formation of methane within the rumen but also it has been found that LCPUFA inhibit the biofilm formation in gram-positive microbes affecting their viability and metabolism. Nevertheless, these metabolic changes should be validated through methane determination on a farm-scale.

In conclusion, algae and marine microbes constitute an enormous category of bioactive compounds with great plasticity making their future on scale application not only promising but also indispensable.

Food Waste as Animal Feed: A Neglected Approach of Great Potential

Every year an abundant amount of 1.3 billion tons of food produced for human consumption, which corresponds to one third of the total production, gets lost or wasted [74]. According to several studies [75], the generation of food waste has a negative impact on the environment. Since food waste has high nutritional value [76] and can be a valuable source of protein, it can be used in animal nutrition. These attributes stem from the fact that food waste such as catering waste may contain a variety of food components, such as plant (e.g., grains, pulses), animal (e.g., meat, eggs, milk) and fish materials. Food waste is characterized by high moisture and salt content, and variability in nutrient composition, which partially depends on the food components that food waste is consisted of [77, 78]. These nutritional hurdles can be mitigated by mixing of food waste with other feed ingredients to formulate a balanced diet [79], and food waste can be fed wet, dried, or ensiled, while direct feeding without processing may jeopardize animal's health due to safety hazards (e.g., pathogens, undesirable materials). Safety concerns are the main reason for the prohibition of food waste in animal diets in the European Union (EU) [80], while it is strictly regulated in the US [81] and in Asian countries like Japan, South Korea, and Taiwan [82]. Food waste has long been used as feed for livestock, especially for pigs [4]. Pigs and poultry are omnivorous species suitable to convert food waste potentially containing animal materials to animal products, such as meat and eggs. Food waste is also recycled as fish feed in countries in Asia [82], while it has been proposed as feed for insects destined for utilization in fish nutrition [83].

The incorporation of food waste in an analysis of 19 different feed trials did not affect pig feed conversion efficiencies and 17 meat quality traits, such as drip loss, flavor, overall palatability, meat color, backfat thickness, dressing percentage, and juiciness [4]. However, growth rates tended to be slightly lower in comparison with swine fed conventional diets. In 6 of the 18 studies tested, the incorporation of 50% food waste in diets resulted in 1.4% higher monounsaturated fatty acids (MUFAs) and 13% greater meat marbling in the produced meat, but no significant differences were observed for PUFA and SFA contents. In one of the aforementioned studies used in the analysis of zu Ermgassen et al. [4], Kjos et al. [84] observed significantly higher DHA and linoleic acid at 40% and 60% inclusion levels, and higher α-linolenic acid content at 20, 40 and 60% inclusion levels in backfat of pigs fed a food waste product containing food leftovers, bakery waste and other types of waste or by-products. At the same time, little or no effect was found on carcass characteristics and meat quality. In monogastrics, the fatty acid profile of the animal products is largely linked to this of the ingested feed [85]. Therefore, the fatty acid profile of meat and fat of pigs can be manipulated based on the feed components of the diet.

In another study, Choe et al. [86] found that total PUFAs, linoleic acid, α-linolenic acid, EPA, DHA, PUFA:SFA ratio, n-3 and n-6 fatty acids levels were significantly higher in pork loin and backfat from pigs fed restaurant waste (RW) in comparison with the control group (0% RW). As PUFAs are highly susceptible to peroxidation, Choe et al. [86] observed increased TBARS values of the pork loin from the group fed RW. Supplementation with antioxidants in the diet, decreased exposure to high heat and oxygen, and limited storage prior to collection and processing of food waste can improve oxidative stability of meat [87].

In broilers, the addition of 10% and 30% of dried leftover food (DLF), 10% DLF and 5% higher protein level (PL), and 20% DLF and 10% higher PL in diets significantly increased the linolenic acid level in the derived meat [88]. Furthermore, significantly higher linoleic acid level was observed in broiler meat from groups fed diets containing 10% DLF, and 10% DLF with 5% higher PL. EPA concentration in meat was also found to be significantly elevated in groups fed diets containing 30% DLF, 30% DLF and 15% higher PL, and 10% DLF and 5% higher PL [88]. Regarding broiler performance and health, inclusion of solar-dried hotel food waste in diets at the level of 15% did not affect FCR, mortality, blood parameters, carcass or breast yield, and biochemical parameters of broilers. However, decreased final body weight and yellowness of meat and elevated serum cholesterol levels in the food-waste-fed compared to the control broilers were observed [89].

In laying hens, incorporation of 50% dehydrated kitchen waste (DKW) in the diet significantly increased the sum of PUFAs and linoleic acid and lowered the sum of SFAs [90]. At the same time, incorporation of DKW at the levels of 12.5% and 25% in the diet did not affect body weight gain and egg weight and increased feed intake, but the same parameters were decreased at 50% DKW inclusion. Although no significant differences were observed for feed conversion, egg production, thickness of albumen, and Haugh units between experimental (12.5, 25, 50% DKW) and control (0% DKW) groups, eggshell strength, redness (a* values) of eggshell, and Roche color fan values decreased and lightness (L* values) of eggshell increased with increasing percentages of DKW [90].

From an environmental perspective, feeding food waste to pigs in the EU to rates akin to those in Japan and South Korea has the potential to reduce land use of pork production by 20% and save 8.8 million tons of grains which could be used for human consumption instead [4]. Furthermore, in several studies it has been suggested that the practice of feeding food waste to pigs can have favorable effect on several environmental impacts such as GHG emissions and eutrophication as reviewed by zu Ermgassen et al. [4], especially when wet feeding takes place instead of dry feeding [91].

Hence, the once traditional practice of feeding food waste to pigs and, by extension, to other animals, provides promising results in terms of animal products' quality, environmental sustainability, and animal performance without compromising animal health. In this way, waste management could be aided without replacing efforts for prevention and re-use of food waste, while simultaneously the environmental footprint of animal production could be alleviated.

Agro-Industrial By-Products

Apart from their nutritional value, agro-industrial by-products contain several secondary compounds that fortify animal products and boost human health. Due to their low cost, they are considered as a cheap source of functional additives – nutraceuticals and their inclusion in animal diets minimizes the environmental implications induced by their disposal (due to their high organic load) while the sustainable use of high-added value ingredients inside food chain is ensured. At the same time, the application of agro-industrial by-products in livestock diets is preferable, since the market demand for clean, natural, or eco-green animal products that are safe and healthy and obtained from sustainable farming systems is high [92]. Agro-industrial by-products could be used either as raw material or after processing, such as drying for the reduction of moisture content or other advanced techniques for the collection of specific compounds well known as nutraceuticals. However, the variation in their composition among different regions and the limited knowledge concerning aspects such as their application (mode, dose etc.), functional pathways, bioactivity, bioavailability, and interactions with other ingredients in animal gastrointestinal tract pose limitations on their regular use [93].

Apple Pomace

Apple pomace (AP) is usually incorporated into pig diets, since it contains desirable nutrients, PUFAs and bioactive substances such as vitamins, anthocyanins, and phenolic acids [94], resulting in the improvement of pig growth performance and meat quality. Dietary addition of fermented AP (2%) seems to improve growth, feed intake, and carcass weight in finishing pigs. At the same time, meat moisture and crude protein contents are decreased, pH_{24} and water holding capacity values are increased and marbling of fresh meat but also tenderness, juiciness, flavor, and overall acceptability of cooked meat are improved [95]. AP supplemented diets (4%) also improved average daily gain and feed conversion in piglets, as assessed 3 weeks after weaning [96].

Furthermore, incorporation of AP (400 or 800 mg/kg) into the diets of finishing pigs improved meat nutritional value. The higher dose also ameliorated carcass traits, meat color, and meat flavor [97]. Improvement of growth parameters and meat antioxidant capacity were also observed after the inclusion of 1% fermented AP in the diets of broilers [98], but higher AP levels leads to a deterioration of growth performance, total antioxidant capacity and blood parameters [99]. In cows, up to 30% of the diet could be efficiently substituted by ensiled mixed tomato and apple pomace leading to a slight increase in milk yield without any other adverse effects [100]. Moreover, inclusion of ensiled mixed apple pomace and broiler litter (15 or 30%) in dairy cows' rations might efficiently result in an improvement on milk yield and dry matter intake [101].

Citrus Pulp

Dried citrus pulp (CP) represents a rich source of energy, fiber and calcium and is a cheap feedstuff for ruminants especially during summer when grass is limited in countries around the Mediterranean. As it has been demonstrated, milk concentration in total polyphenols and flavonoids and milk antioxidant capacity were increased after feeding dairy cattle with pelleted CP (9-18%) [102]. Replacement of cereals by dried CP in concentrated-based diets (24-35%) could improve meat oxidative [103, 104] and protein [105] stability and increase intramuscular PUFAs concentration [106] in sheep. Replacement of barley by dried CP (40-80%) [107] or CP dietary inclusion at the level of (150 g/kg) [108] improved fatty acid profile of beef without affecting meat oxidative and color stability. Lower CP levels (15%) do not appear to influence PUFAs and SFAs levels [109]. Moreover, no effect of dietary CP supplementation at different levels (10-45%) [110-112] on feed intake, growth performance and carcass characteristics in lambs, is observed.

CP inclusion at the level of 9% in the diets of laying hens improved egg yolk oxidative stability, but impairment of performance and egg quality were observed [113]. In broilers, incorporation of CP in the diet up to the level of 10% did not have an effect on body weight, carcass yield, intestinal morphometry, and meat quality; although a delayed and reduced oxidation rate in chicken meat was observed [114]. In ostrich, CP inclusion (20%) in the diet reduced meat cooking loss, increased PUFA content and n-6/n-3 ratio [115]. No effect of dried CP up to the level of 12% on growth performance and carcass yield was found in geese [116]. Including 7.5% of dried CP to the diet did not compromise overall growth performance, but slightly increased damage in the gut morphology of weanling pigs [117]. Replacing 15% of the dietary dry matter with dried CP did not have a significant effect on carcass traits, meat quality or sensory characteristics in pigs [118]. The inclusion of ensiled CP at the level of 5-10% in diets for growing pigs did not also affect back-fat thickness, color parameters and the proportion and composition of SFAs and PUFAs in pork meat [119]. On the other hand, Watanabe et al. [120] found a negative linear effect of dietary supplementation with CP (10-30%) on final body weight, carcass yield and pork meat color characteristics.

Grape Pomace

Recent research has stressed the importance of by-products from wine processing as plant materials that could be used in animal and particularly broiler production [121]. Dietary supplementation with grape pomace (GP) at levels of 5-60 g/kg in broilers increased the antioxidant capacity of breast muscle and PUFAs content, improved intestinal morphology and decreased serum cholesterol levels although growth performance, feed intake, feed efficiency and weights of abdominal fat, liver, pancreas and spleen were not affected [121-126]. Dietary GP supplementation may also have a beneficial impact on piglets' welfare and may improve productivity as well as meat quality [127]. In pigs, GP dietary inclusion (30 g/kg) fermented by *Saccharomyces boulardii* increased marbling score, values of redness (a*) and yellowness (b*) and the anti-oxidative ability (lower TBARS) of pork meat [128]. Addition of ensiled GP (3.5-7%) resulted in meat with increased PUFA/SFA ratio and reduced n-6/n-3 ratio [129]. GP supplementation (5%) appeared to improve kidneys, liver and spleen antioxidant status in piglets [130]. In lambs, utilization of GP up to 10% in diet did not have negative effects on growth performance, carcass traits and meat quality, while antioxidant status, meat n-6 levels and oxidative stability were improved [131-135]. Grape pomace (GP) could also induce an improved meat oxidative stability compared to that of CP when included in beef cattle finishing diets (150g/kg) [136].

Olive Cake

One feasible application of olive by-products (OB) is their incorporation in the diets of livestock and especially ruminants due to their high fiber content. As indicated by numerous studies, OB dietary supplementation increases the levels of MUFAs and decreases that of SFAs in milk and meat of ruminants with beneficial effects for consumers' health. At the same time, environmental impact and feeding costs are reduced without detrimental effects on ruminal fermentation, nutrients utilization, growth performance, carcass traits, milk yield and composition [137]. Furthermore, addition of olive cake (OC) in pig diets could serve as an alternative strategy of reducing production costs, especially in the Mediterranean region, while the quality and antioxidant capacity of the derived meat is maintained [138], as indicated by the reduction in the levels of SFAs and the increase in that of unsaturated fatty acids (MUFAs and PUFAs) [139]. In broilers, meat MUFAs levels are increased [140] and growth parameters [141] and carcass traits [142] are improved as an effect of OC dietary inclusion. Similar findings have been shown in laying hens, since levels of cholesterol and SFAs are reduced and that of MUFAs and PUFAs are increased in egg yolk after OC dietary supplementation, without negative effects on productive performance at a level up to 16% [143].

Pomegranate By-Products

Addition of dried pomegranate by-products (PB) (50-150 g/kg) linearly increased fat content and decreased shear force, drip loss, total aerobic bacterial count, and lipid oxidation of longissimus lumborum muscle in kids. At the same time, a linear increase of linoleic acid, α-linolenic acid, n-6, and n-3 PUFA and a decrease in the ratio of n-6/n-3 in both muscle and adipose tissues, and a linear increase in vaccenic acid, CLA and punicic acid concentration in subcutaneous and intramuscular fat were observed [144]. In lambs, dietary PB supplementation improved meat quality characteristics and antioxidant potential, as indicated by the increase in PUFA and essential fatty acids, linoleic, α-linolenic, and trans-10, cis-12 CLA acids in intramuscular fat and the increase in total phenolic content and antioxidant activity [145-147]. In broilers, PB dietary supplementation (0.5-2%) also reduced lipid oxidation values and SFA and increased MUFA, PUFA and n-3 fatty acids in meat [148]. Furthermore, addition of PB (2-8%) may improve performance and activity of meat antioxidant enzyme biomarkers [149] or not [150] in broilers.

Tomato Pomace

Tomato pomace (TP) can also be fed to livestock, fresh, dried, or ensiled with or without additives as an alternative cheap source of energy and other nutrients. Replacement of maize silage with ensiled TP (10%) had a positive effect on milk vitamin contents, antioxidant, and immune functions in the early lactating dairy cows [151]. Inclusion of TP (40%) improves both milk production and composition without affecting the physiological characteristics of goats [152]. At the same time, amounts of TP up to 180 g/kg could be included in a diet for fattening ruminants without negatively affecting rumen fermentation [153, 154]. TP (5%) applied as fibre source in weaned piglets did not improve color, fatty acid profile, cholesterol content or increased levels of lycopene or β-carotene in meat. However, a positive increase in α-tocopherol content of meat and liver was observed indicating an improvement in pork oxidative stability [155]. TP dietary inclusion reduced intramuscular fat, SFA and MUFA content, while increased PUFA and n-6:n-3 ratio in meat [156]. In fattening rabbits, TP dietary addition at the level of 6% resulted in increased PUFA levels, yellowness (b*) and chroma values [157]. Increase of PUFA and of PUFA/SFA ratio was also observed in the meat of quails fed with a dried tomato pulp supplemented diet at the level of 10%. However, effects on the oxidative stability of meat were not exerted at this concentration but at 5% [158]. Sahin et al. [159] reached to the same conclusions; MDA levels in serum, liver and muscle of quails decreased after dietary TP supplementation at the levels of 2.5-5%.

Conclusion

As the world population increases along with the demand for animal products and while there is great interest in foods containing health-protective substances, efforts to implement novel tools to address these challenges should be robustly promoted by all stakeholders through collaborative and multidisciplinary actions, most notably, between the feed and food industry and researchers. Alternative feedstuffs contain bioactive compounds that improve the oxidative stability and thus extend the shelf-life of the derived animal products, which also have health-promoting properties. These valuable components can improve animal health and reduce the dependence on antibiotics in animals. Additionally, increased awareness for environmental challenges has shifted the direction of animal production to implement more environmentally friendly practices. In order to correspond to the current trends, the utilization of alternative feedstuffs in animal nutrition is an appealing solution.

References

[1] United Nations, Department of Economic and Social A airs Population Division. *World Population Prospects 2019: Highlights.* New York: United Nations, Department of Economic and Social A airs Population Division; 2019.

[2] Alexandratos, N, Bruinsma, J. *World Agriculture Towards 2030/2050: The 2012 Revision.* Rome: Food and Agriculture Organization of the United Nations, Agricultural Development Economics Division (ESA); 2012.

[3] Nepstad, DC, Stickler, CM, Almeida, OT. Globalization of the Amazon Soy and Beef Industries: Opportunities for Conservation. *Conserv Biol* 2006;20:1595-1603.

[4] zu Ermgassen, EKHJ, Phalan, B, Green, RE, Balmford, A. Reducing the land use of EU pork production: Where there's swill, there's a way. *Food Policy* 2016;58:35-48.

[5] Gerber, PJ, Steinfeld, H, Henderson, B, Mottet, A, Opio, C, Dijkman, J, et al. *Tackling climate change through livestock – A global assessment of emissions and mitigation opportunities.* Rome: Food and Agriculture Organization of the United Nations (FAO); 2013.

[6] Maneerat, W, Prasanpanich, S, Tumwasorn, S, Laudadio, V, Tufarelli, V. Evaluating agro-industrial by-products as dietary roughage source on growth performance of fattening steers. *Saudi J Biol Sci* 2015;22(5):580-584.

[7] Waseem, S, Imadi, SR, Gul, A, Ahmad, P. Oilseed crops: Present scenario and future prospects. In Ahmad, P. (Ed.), *Oilseed Crops* (pp. 1–18). Chichester: John Wiley & Sons Ltd; 2017.

[8] Bernard, JK. Oilseed and Oilseed Meals. In *Reference Module in Food Science* (pp. 349-355). Elsevier; 2016.

[9] Tanwar, B, Goyal, A. (2021). *Oilseeds: Health Attributes and Food Applications.*

[10] Schmidt, Š, Pokorný, J. Potential Application of Oilseeds as Sources of Antioxidants for Food Lipids - A Review. *Czech J Food Sci* 2005;23(3): 93–102.

[11] Dunford, NT. Oil-and oilseed-based bioactive compounds and their health effects. In: Dunford, NT, Dunford, HB (Eds.). *Nutritionally enhanced edible oil and oilseed processing* (pp. 19-23). Champaign: AOCS Press; 2004.

[12] McKevith, B. Nutritional Aspects of Oilseeds. *Nutr. Bull.* 2005;30(1):13–26.

[13] Cromwell GL. *Soyabean Meal: An Exceptional Protein Source.* Soyabean Meal InfoCenter. 2012. Available at: https://www.soymeal.org/wp-content/uploads/2018/04/soybean_meal_an_exceptional_protein_source.pdf.

[14] Charron, CS, Allen, FL, Johnson, RD, Pantalone, VR, Sams, CE. Correlations of Oil and Protein with Isoflavone Concentration in Soybean [Glycine Max (L.) Merr.]. *J Agric Food Chem* 2005; 53(18):7128–7135.

[15] Natarajan, S, Luthria, D, Bae, H, Lakshman, D, Mitra, A. Transgenic Soybeans and Soybean Protein Analysis: An Overview. *J Agric Food Chem* 2013;61(48):11736–11743.

[16] Modgil, R, Tanwar, B, Goyal, A, Kumar, V. Soybean (Glycine max). In: Tanwar B, Goyal A (Eds.) *Oilseeds: Health Attributes and Food Applications.* Singapore: Springer; 2021.

[17] Kliem, KE, Shingfield, KJ. Manipulation of Milk Fatty Acid Composition in Lactating Cows: Opportunities and Challenges. *Eur J Lipid Sci Technol* 2016;118(11):1661–1683.

[18] Zhang, RH, Mustafa, AD, Zhao, X. Effects of Feeding Oilseeds Rich in Linoleic and Linolenic Fatty Acids to Lactating Ewes on Cheese Yield and on Fatty Acid Composition of Milk and Cheese. *Anim Feed Sci Technol 2006*;127(3–4):220–233.

[19] Nudda, A, Battacone, G, Fancellu, S, Carboni, GA, Pulina, G. The Use of Linseed and Cottonseed to Change the Milk Fatty Acid Profile in Early Lactation Dairy Goats. In Priolo, A, Biondi, L, Ben Salem, H, Morand-Fehr, P (Eds.). *Advanced nutrition and feeding strategies to improve sheep and goat* (pp. 49-53). Zaragoza: CIHEAM; 2007. (Options Mediterraneennes. Serie A, Seminaires Mediterraneens).

[20] Mitsiopoulou, C, Sotirakoglou, K, Labrou, NE, Tsiplakou, E. The Effect of Whole Sesame Seeds on Milk Chemical Composition, Fatty Acid Profile and Antioxidant Status in Goats. *Livest Sci* 2021;245:104452.

[21] Kliem, KE, Humphries, DJ, Kirton, P, Givens, DI, Reynolds, CK. Differential Effects of Oilseed Supplements on Methane Production and Milk Fatty Acid Concentrations in Dairy Cows. *Animal* 2019;13(2):309–17.

[22] Hurtaud, C, Peyraud, JL. Effects of Feeding Camelina (Seeds or Meal) on Milk Fatty Acid Composition and Butter Spreadability. *J Dairy Sci* 2007;90(11):5134–5145.

[23] Goiri, I, Zubiria, I, Benhissi, H, Atxaerandio, R, Ruiz, R, Mandaluniz, N, Garcia-Rodriguez, A. Use of Cold-Pressed Sunflower Cake in the Concentrate as a Low-Input Local Strategy to Modify the Milk Fatty Acid Profile of Dairy Cows. *Animals* 2019;9(10):803.

[24] Amores, G, Mailo, V, Nájera, AI, Mandaluniz, N, Arranz, J, Bustamante, MA, et al. Rapeseed and Sunflower Oilcake as Supplements for Dairy Sheep: Animal Performance and Milk Fatty Acid Concentrations. *J Dairy Res* 2014;81(4):410–416.

[25] Jóźwik, A, Strzałkowska, N, Bagnicka, E, Łagodziński, Z, Pyzel, B, Chyliński, W, et al. The Effect of Feeding Linseed Cake on Milk Yield and Milk Fatty Acid Profile in Goats. *Anim Sci Pap Rep* 2010;28(3):245–251.

[26] Mitsiopoulou, C, Karaiskou, C, Simoni, M, Righi, F, Pappas, AC, Sotirakoglou, K, Tsiplakou, E. Influence of Dietary Sesame Meal, Vitamin E and Selenium Supplementation on Milk Production, Composition, and Fatty Acid Profile in Dairy Goats. *Livest Sci* 2020;244:104336.

[27] Zając, M, Kiczorowska, B, Samolińska, W, Kowalczyk-Pecka, D, Andrejko, D, Kiczorowski, P. Effect of Inclusion of Micronized Camelina, Sunflower, and Flax Seeds in the Broiler Chicken Diet on Performance Productivity, Nutrient Utilization, and Intestinal Microbial Populations. *Poult Sci* 2021;101118.

[28] McDonnell, P, O'Shea, C, Figat, S, O'Doherty JV. Influence of Incrementally Substituting Dietary Soya Bean Meal for Rapeseed Meal on Nutrient Digestibility, Nitrogen Excretion, Growth

Performance and Ammonia Emissions from Growing-Finishing Pigs. *Arch Anim Nutr* 2010;64(5):412–424.

[29] Rahimi, S, Azad, SK, Torshizi, MAK. Omega-3 Enrichment of Broiler Meat by Using Two Oil Seeds. *J Agric Sci Technol* 2011;13:353-365.

[30] Cherian, G, Campbell, A, Parker, T. Egg Quality and Lipid Composition of Eggs from Hens Fed Camelina Sativa. *J Appl Poult Res* 2009;18(2):143–150.

[31] Aziza, AE, Quezada, N, Cherian, G. Feeding Camelina Sativa Meal to Meat-Type Chickens: Effect on Production Performance and Tissue Fatty Acid Composition. *J Appl Poult Res* 2010;19(2): 157–168.

[32] Aziza, AE, Quezada, N, Cherian, G. Antioxidative Effect of Dietary Camelina Meal in Fresh, Stored, or Cooked Broiler Chicken Meat. *Poult Sci* 2010;89(12):2711–2718.

[33] Tsiplakou, E, Mitsiopoulou, C, Karaiskou, C, Simoni, M, Pappas, AC, Righi F, et al. Sesame Meal, Vitamin E and Selenium Influence Goats' Antioxidant Status. *Antioxidants* 2021;10(3):392.

[34] Schogor, A, Palin, M, Dos Santos, G, Benchaar, C, Lacasse, P, Petit, H. Mammary Gene Expression and Activity of Antioxidant Enzymes and Oxidative Indicators in the Blood, Milk, Mammary Tissue and Ruminal Fluid of Dairy Cows Fed Flax Meal. *Br J Nutr* 2013;110(10):1743–1750.

[35] Côrtes, C, Palin, M-F, Gagnon, N, Benchaar, C, Lacasse, P, Petit, HV. Mammary Gene Expression and Activity of Antioxidant Enzymes and Concentration of the Mammalian Lignan Enterolactone in Milk and Plasma of Dairy Cows Fed Flax Lignans and Infused with Flax Oil in the Abomasum. *Br J Nutr* 2012; 108(8):1390–1398.

[36] Mitsiopoulou, C, Sotirakoglou, K, Skliros, D, Flemetakis, E, Tsiplakou, E. The Impact of Whole Sesame Seeds on the Expression of Key-Genes Involved in the Innate Immunity of Dairy Goats. *Animals* 2021;11(2):468.

[37] Salehi, R, Colazo, MG, Gobikrushanth, M, Basu, U, Ambrose, D J. Effects of Prepartum Oilseed Supplements on Subclinical Endometritis, pro- and Anti-Inflammatory Cytokine Transcripts in Endometrial Cells and Postpartum Ovarian Function in Dairy Cows. *Reprod Fertil Dev* 2016;29(4):747–758.

[38] Zhan, ZP, Huang, FR, Luo, J, Dai, JJ, Yan, XH, Peng, J. Duration of Feeding Linseed Diet Influences Expression of Inflammation-Related Genes and Growth Performance of Growing-Finishing Barrows. *J Anim Sci* 2009;87(2):603–611.

[39] Alvarez-Hess, PS, Williams, SR O, Jacobs, JL, Hannah, MC, Beauchemin, KA, Eckard, R J, et al. Effect of Dietary Fat Supplementation on Methane Emissions from Dairy Cows Fed Wheat or Corn. *J Dairy Sci* 2019;102(3):2714–2723.

[40] Beauchemin, KA, McGinn, SM, Benchaar, C, Holtshausen, L. Crushed Sunflower, Flax, or Canola Seeds in Lactating Dairy Cow Diets: Effects on Methane Production, Rumen Fermentation, and Milk Production. *J Dairy Sci* 2009;92(5):2118–2127.

[41] Chuntrakort, P, Otsuka, M, Hayashi, K, Takenaka, A, Udchachon, S, Sommart, K. The Effect of Dietary Coconut Kernels, Whole Cottonseeds and Sunflower Seeds on the Intake, Digestibility and Enteric Methane Emissions of Zebu Beef Cattle Fed Rice Straw Based Diets. *Livest Sci* 2014;161(1):80–89.

[42] Abuelfatah, K, Zuki, AB, Goh, YM, Sazili, AQ, Abubakr, A. Effects of Feeding Whole Linseed on Ruminal Fatty Acid Composition and Microbial Population in Goats. *Anim Nutr* 2016; 2:323–28.

[43] Bayat, AR, Kairenius, P, Stefański, T, Leskinen, H, Comtet-Marre, S, Forano, E et al. Effect of Camelina Oil or Live Yeasts (Saccharomyces Cerevisiae) on Ruminal Methane Production, Rumen Fermentation, And Milk Fatty Acid Composition in Lactating Cows Fed Grass Silage Diets. *J Dairy Sci* 2015;98(5): 3166–3181.

[44] Vargas, J, Andrés, S, Ruiz, DY, López, S. The Effect of Olive, Sunflower or Linseed Oils on the Fermentation Pattern and

Methane Production in the Rumen Simulating Technique. *Options Méditerr A* 2011;99:163–168.

[45] Morais, T, Inácio, A, Coutinho, T, Ministro, M, Cotas, J, Pereira, L., Bahcevandziev, K. Seaweed Potential in the Animal Feed: A Review. *J Mar Sci Eng* 2020;8:559.

[46] Madeira, MS, Cardoso, C, Lopes, PA, Coelho, D, Afonso, C, Bandarra, N M, Prates, J A. Microalgae as feed ingredients for livestock production and meat quality: A review. *Livest Sci* 2017;205:111–121.

[47] Benemann, J. Microalgae for Biofuels and Animal Feeds. *Energies* 2013;6:5869-5886.

[48] Remize, M, Brunel, Y, Silva, JL, Berthon, J-Y, Filaire, E. Microalgae n-3 PUFAs Production and Use in Food and Feed Industries. *Mar Drugs* 2021;19(2):113.

[49] Martins, DA, Custódio, L, Barreira, L, Pereira, H, Ben-Hamadou, R, Varela, J, Abu-Salah, K M. Alternative sources of n-3 long-chain polyunsaturated fatty acids in marine microalgae. *Mar Drugs* 2013;11(7):2259-2281.

[50] Xin, Y, Shen, C, She, Y, Chen, H, Wang, C, Wei, L, et al. Biosynthesis of Triacylglycerol Molecules with a Tailored PUFA Profile in Industrial Microalgae. *Mol Plant* 2019;12(4):474-488.

[51] Patelou, M, Infante, C, Dardelle, F, Randewig, D, Kouri, ED, Udvardi, MK, et al. Transcriptomic and metabolomic adaptation of Nannochloropsis gaditana grown under different light regimes. *Algal Res.* 2020;45:101735.

[52] Bruneel, C, Lemahieu, C, Fraeye, I, Ryckebosch, E, Muylaert, K, Buyse, J, Foubert, I. Impact of microalgal feed supplementation on omega-3 fatty acid enrichment of hen eggs. *J Funct Foods* 2013;5(2):897–904.

[53] Petrolli, TG, Petrolli, OJ, Pereira, ASC, Zotti, CA, Romani, J, Villani, R, et al. Effects of the dietary supplementation with a microalga extract on broiler performance and fatty-acid meat profile. *Braz J Poult Sci* 2019;21:3.

[54] Franklin, S, Martin, K, Baer, R, Schingoethe, D, Hippen, A. Dietary Marine Algae (Schizochytrium sp.) Increases Concentrations of Conjugated Linoleic, Docosahexaenoic and Transvaccenic Acids in Milk of Dairy Cows. *J Nutr* 1999;12(11): 2048-2054.

[55] Bichi, E, Hervas, G, Toral, PG, Loor, JJ, Frutos, P. Milk fat depression induced by dietary marine algae in dairy ewes: persistency of milk fatty acid composition and animal performance responses. *J Dairy Sci* 2013;96: 524-532.

[56] Fougère, H, Delavaud, C, Bernard, L. Diets supplemented with starch and corn oil, marine algae, or hydrogenated palm oil differentially modulate milk fat secretion and composition in cows and goats: A comparative study *J Dairy Sci* 2018;101(9):8429-8445.

[57] Mavrommatis, A, Tsiplakou, E. The impact of the dietary supplementation level with Schizochytrium sp. on milk chemical composition and fatty acid profile of both blood plasma and milk of goats. *Small Rumin Res* 2020;193:106252.

[58] Mavrommatis, A, Sotirakoglou, K, Skliros, D, Flemetakis, E, Tsiplakou, E. Dose and time response of dietary supplementation with Schizochytrium sp. on the abundances of several microorganisms in the rumen liquid of dairy goats. *Livest Sci* 2021;247:104489.

[59] Mavrommatis, A, Skliros, D, Simoni, M, Righi, F, Flemetakis, E, Tsiplakou, E. Alterations in the Rumen Particle-Associated Microbiota of Goats in Response to Dietary Supplementation Levels of Schizochytrium spp. *Sustainability* 2021;13:607.

[60] Moran, CA, Morlacchini, M, Keegan, JD, Delles, R, Fusconi, G. Effects of a DHA-rich unextracted microalgae as a dietary supplement on performance, carcass traits and meat fatty acid profile in growing-finishing pigs. *J Anim Physiol Anim Nutr* 2018;102(4):1026–1038.

[61] De Tonnac, A, Mourot, J. Effect of dietary sources of n-3 fatty acids on pig performance and technological, nutritional and sensory qualities of pork. *Animal* 2017;12(7):1527–1535.

[62] Mavrommatis, A, Chronopoulou, G, Sotirakoglou, K, Labrou, N, Zervas, G, Tsiplakou, E. The impact of the dietary supplementation level with Schizochytrium sp, on the oxidative capacity of both goats' organism and milk. *Livest Sci* 2018;218:37–43.

[63] Zhu La, ALT, Pierce, KM, Liu, WH, Gao, ST, Bu, DP, Ma, L. Supplementation with Schizochytrium sp. enhances growth performance and antioxidant capability of dairy calves before weaning. *Anim Feed Sci Technol* 2021;271:114779.

[64] Tsiplakou, E, Abdullah, MAM, Mavrommatis, A, Chatzikonstantinou, M, Skliros, D, Sotirakoglou, K, et al. The effect of dietary Chlorella vulgaris inclusion on goat's milk chemical composition, fatty acids profile and enzymes activities related to oxidation. *J Anim Physiol Anim Nutr* 2017;102(1):142–151.

[65] Tsiplakou, E, Abdullah, MAM, Mavrommatis, A, Chatzikonstantinou, M, Skliros, D, Sotirakoglou, K, et al. The effect of dietary Chlorella pyrenoidosa inclusion on goats milk chemical composition, fatty acids profile and enzymes activities related to oxidation. *Livest Sci* 2017;197:106–111.

[66] Amaro, HM, Guedes, AC, Malcata, FX. Antimicrobial activities of microalgae: An invited review. In Méndez-Vilas, A (Ed.). *Science against Microbial Pathogens: Communicating Current Research and Technological Advances* (pp. 1272-1280). Badajoz: Formatex Research Center; 2011.

[67] El-Bahr, S, Shousha, S, Shehab, A, Khattab, W, Ahmed-Farid, O, Sabike, I, et al. Effect of Dietary Microalgae on Growth Performance, Profiles of Amino and Fatty Acids, Antioxidant Status, and Meat Quality of Broiler Chickens. *Animals* 2020;10(5):761.

[68] Mavrommatis, A, Giamouri, E, Tavrizelou, S, Zacharioudaki, M, Danezis, G, Simitzis, P E, et al. Impact of Mycotoxins on Animals' Oxidative Status. *Antioxidants* 2021;10(2):214.

[69] Bortoluzzi, C, Schmidt, JM, Bordignon, HLF, Fülber, LM, Layter, J R, Fernandes, JIM. Efficacy of yeast derived glucomannan or algae-based antioxidant or both as feed additives to ameliorate mycotoxicosis in heat stressed and unstressed broiler chickens. *Livest Sci* 2016;193:20–25.

[70] Caroprese, M, Albenzio, M, Ciliberti, MG, Francavilla, M, Sevi, A. A mixture of phytosterols from Dunaliella tertiolecta affects proliferation of peripheral blood mononuclear cells and cytokine production in sheep. *Vet Immunol Immunopathol* 2012;150:27–35.

[71] Ciliberti, MG, Albenzio, M, Francavilla, M, Neglia, G, Esposito, L, Caroprese, M. Extracts from Microalga Chlorella sorokiniana Exert an Anti-Proliferative Effect and Modulate Cytokines in Sheep Peripheral Blood Mononuclear Cells. *Animals* 2019;9(2):45.

[72] Roque, BM, Brooke, CG, Ladau, J, Polley, T, Marsh, LJ, Najafi, N, et al. Effect of the macroalgae *Asparagopsis taxiformis* on methane production and rumen microbiome assemblage. *Anim Microbiome* 2019;1:3.

[73] Roque, BM, Salwen, JK, Kinley, R, Kebreab, E. Inclusion of Asparagopsis armata in lactating dairy cows' diet reduces enteric methane emission by over 50 percent. *J Clean Prod* 2019;234:132–138.

[74] Gustavsson, J, Cederberg, C, Sonesson, U, Otterdijk, R, McYbeck, A. *Global Food Losses and Food Waste: Extent, Causes and Prevention.* Rome: Food and Agriculture Organization of the United Nations; 2011.

[75] Lee, SH, Choi, K, Osako, M, Dong, J. Evaluation of Environmental Burdens Caused by Changes of Food Waste Management Systems in Seoul, Korea. *Sci Total Environ* 2007;378(1):42–53.

[76] Westendorf, ML. Food Waste as Swine Feed. In Westendorf, ML (Ed.), *Food Waste to Animal Feed* (pp. 69-90). Iowa: Iowa State University Press; 2000.

[77] Georganas, A, Giamouri, E, Pappas, AC, Papadomichelakis, G, Galliou, F, Manios, T, et al. Bioactive compounds in food waste: A review on the transformation of food waste to animal feed. *Foods* 2020;9(3):291.

[78] Westendorf, ML. Food Waste as Animal Feed: An Introduction. In Westendorf, ML (Ed.), *Food Waste to Animal Feed* (pp. 3-16). Iowa: Iowa State University Press; 2000.

[79] Kwak, WS, Kang, JS. Effect of feeding food waste-broiler litter and bakery by-product mixture to pigs. *Bioresour Technol* 2006;97(2):243-249.

[80] EC. Regulation (EC) No 1069/2009 of the European Parliament and of the Council of 21 October 2009 laying down health rules as regards animal by-products and derived products not intended for human consumption and repealing Regulation (EC) No 1774/2002 (Animal by-products Regulation). Brussels: EC; 2009.

[81] Leib, EB, Balkus, O, Rice, C, Maley, M, Taneja, R, Cheng, R, et al. *Leftovers for Livestock: A Legal Guide for Using Excess Food as Animal Feed.* Fayetteville: The Harvard Food Law and Policy Clinic and the Food Recovery Project at the University of Arkansas School of Law; 2016.

[82] Cheng, JYK, Irene MCLo. Investigation of the available technologies and their feasibility for the conversion of food waste into fish feed in Hong Kong. *Environ Sci Pollut Res* 2016;23(8): 7169-7177.

[83] Stamer, A. Insect proteins-a new source for animal feed: The use of insect larvae to recycle food waste in high-quality protein for livestock and aquaculture feeds is held back largely owing to regulatory hurdles. *EMBO Rep.* 2015;16(6):676-680.

[84] Kjos, NP, Øverland, M, Arnkværn Bryhni, E, SØrheim, O. Food Waste Products in Diets for Growing-finishing Pigs: Effect on

Growth Performance, Carcass Characteristics and Meat Quality. *Acta Agric Scand A Anim Sci* 2000;50(3):193-204.

[85] Kouba, M, Mourot, J. A review of nutritional effects on fat composition of animal products with special emphasis on n-3 polyunsaturated fatty acids. *Biochimie* 2011;93(1):13-17.

[86] Choe, J, Moyo, KM, Park, K, Jeong, J, Kim, H, Ryu, Y, et al. Meat Quality Traits of Pigs Finished on Food Waste. *Korean J Food Sci Anim Resour* 2017;37(5):690-697.

[87] Fung, L, Urriola, PE, Baker, L, Shurson, GC. Estimated energy and nutrient composition of different sources of food waste and their potential for use in sustainable swine feeding programs. *Transl Anim Sci* 2018;3(1):359-368.

[88] Cho, YM, Lee, GW, Jang, JS, Shin, IS, Myung, KH, Choi, KS, et al. Effects of feeding dried leftover food on growth and body composition of broiler chicks. Asian-australas. *J Anim Sci* 2004;17(3):386-393.

[89] Giamouri, E, Pappas, AC, Papadomichelakis, G, Tsiplakou, E, Sotirakoglou, K, Markakis, N, et al. The food for feed concept. Performance of broilers fed hotel food residues. *Br Poult Sci* 2021;62(3):452-458.

[90] Kojima, S. Dehydrated kitchen waste as a feedstuff for laying hens. *Int J Poult Sci* 2005;4(9):689-694.

[91] Salemdeeb, R, zu Ermgassen, EKHJ, Kim, MH, Balmford, A, Al-Tabbaa, A. Environmental and health impacts of using food waste as animal feed: A comparative analysis of food waste management options. *J Clean Prod* 2017;140:871-880.

[92] Font i Furnols, M, Realini, C, Montossi, F, Sanudo, C, Campo, MM, Oliver, MA, et al. Consumer's purchasing intention for lamb meat affected by country of origin, feeding system and meat price: A conjoint study in Spain, France and United Kingdom. *Food Qual Pref* 2011;22:443–451.

[93] Galanakis, C. M. Recovery of high added-value components from food wastes: Conventional, emerging technologies and

commercialized applications. *Trends Food Sci Technol* 2012;26(2): 68-87.

[94] Pieszka, M, Gogol, P, Pietras, M, Pieszka, M. Valuable components of dried pomaces of chokeberry, black currant, strawberry, apple and carrot as a source of natural antioxidants and nutraceuticals in the animal diet. *Ann Anim Sci* 2015;15: 475–491.

[95] Lee, SD, Kim, HY, Jung, HJ, Ji, SY, Chowdappa, R, Ha, JH, et al. The effect of fermented apple diet supplementation on the growth performance and meat quality in finishing pigs. *Anim Sci J* 2009;80:79-84.

[96] Dufourny, S, Antoine, N, Pitchugina, E, Delcenserie, V, Godbout, S, Douny, C, et al. Apple Pomace and Performance, Intestinal Morphology and Microbiota of Weaned Piglets—A Weaning Strategy for Gut Health?. *Microorganisms* 2021;9(3):572.

[97] Xu, X, Chen, X, Chen, D, Yu, B, Yin, J, Huang, Z. Effects of dietary apple polyphenol supplementation on carcass traits, meat quality, muscle amino acid and fatty acid composition in finishing pigs. *Food Funct* 2019;10(11):7426-7434.

[98] Kang, HK, Choi, HC, Kang, BS, Na, JC, Yu, DJ, Kang, GH, et al. Effects of dietary fermented apple pomace and cinnamon addition on meat quality and performance in broiler. *J Anim Sci Technol (Kor)* 2009;51:315-320.

[99] Hosein Aghili, A, Toghyani, M, Ali Tabeidian, S. Effect of incremental levels of apple pomace and multi enzyme on performance, immune response, gut development and blood biochemical parameters of broiler chickens. *Int J Recycl Org Waste Agric* 2019;8:321-334.

[100] Abdollahzadeh, F, Pirmohammadi, R, Fatehi, F, Bernousi, I. Effect of feeding ensiled mixed tomato and apple pomace on performance of Holstein dairy cows. *Slovak J Anim Sci* 2010;43:31–35.

[101] Azizi, O, Karimi, S, Sadeghi, G, Lashkari, S. Milk yield and composition, dry matter intake and blood parameters of Holstein cows fed ensiled apple pomace co-ensiled with broiler litter. *Span J Agric Res,* 2014;12:462-466.

[102] Santos, GT, Lima, LS, Schogor, ALB, Romero, JV, De Marchi, FE, Grande, PA, et al. Citrus pulp as a dietary source of antioxidants for lactating Holstein cows fed highly polyunsaturated fatty acid diets. Asian-Australas *J Anim Sci* 2014;27:1104-1113.

[103] Inserra L, Priolo, A, Biondi, L, Lanza, M, Bognanno, M, Gravador, R, Luciano, G. Dietary citrus pulp reduces lipid oxidation in lamb meat. *Meat Sci* 2014;96:1489-1493.

[104] Luciano, G, Roscini, V, Mattioli, S, Ruggeri, S, Gravador, RS, Natalello, A, et al. Vitamin E is the major contributor to the antioxidant capacity in lambs fed whole dried citrus pulp. *Animal* 2017;11(3),411-417.

[105] Gravador, RS, Jongberg, S, Andersen ML, Luciano, G, Priolo A, Lund, MN. Dietary citrus pulp improves protein stability in lamb meat stored under aerobic conditions. *Meat Sci* 2014;97:231-236.

[106] Lanza, M, Scerra, M, Bognanno, M, Buccioni, A, Cilione, C, Biondi, L, et al.. Fatty acid metabolism in lambs fed citrus pulp. *J Anim Sci* 2015; 93(6):3179-3188.

[107] Salami, SA, O'Grady, MN, Luciano, G, Priolo, A, McGee, M, Moloney, AP, Kerry, JP. Quality indices and sensory attributes of beef from steers offered grass silage and a concentrate supplemented with dried citrus pulp. *Meat Sci* 2020;168:108181.

[108] Tayengwa, T, Chikwanha, OC, Neethling, J, Dugan, ME, Mutsvangwa, T, Mapiye, C. Polyunsaturated fatty acid, volatile and sensory profiles of beef from steers fed citrus pulp or grape pomace. *Food Res Int* 2021;139:109923.

[109] Francisco, A, Alves, SP, Portugal, PV, Dentinho, MT, Jerónimo, E, Sengo, S, et al. Effects of dietary inclusion of citrus pulp and rockrose soft stems and leaves on lamb meat quality and fatty acid composition. *Animal* 2018;12(4):872-881.

[110] Caparra, P, Foti, F, Scerra, M, Sinatra, MC, Scerra, V. Solar-dried citrus pulp as an alternative energy source in lamb diets: effects on growth and carcass and meat quality. *Small Rumin Res* 2007;68:303–311.

[111] Gobindram, MNNE, Bognanno, M, Luciano, G, Avondo, M, Piccione, G, Biondi, L. The effects of barley replacement by dehydrated citrus pulp on feed intake, performance, feeding behaviour and serum metabolic indicators in lambs. *Anim Prod Sci* 2017;57(1):133-140.

[112] Sharif, M, Ashraf, MS, Mushtaq, N, Nawaz, H, Mustafa, MI, Ahmad, F, et al. Influence of varying levels of dried citrus pulp on nutrient intake, growth performance and economic efficiency in lambs. *J Appl Anim Res* 2018;46(1):264-268.

[113] Goliomytis, M, Kostaki, A, Avgoulas, G, Lantzouraki, D Z, Siapi, E, Zoumpoulakis, P, et al. Dietary supplementation with orange pulp (Citrus sinensis) improves egg yolk oxidative stability in laying hens. *Anim Feed Sci Technol* 2018;244:28-35.

[114] Diaz-Vargas, M, Murakami, AE, Pintro, PTM, Ospina-Rojas, IC, de Souza, CHP, Eyng, C. Dehydrated citrus pulp in broiler diets. *Can J Anim Sci* 2018;99(1):33-40.

[115] Lanza, M, Fasone, V, Galofaro, V, Barbagallo, D, Bella, M, Pennisi, P. Citrus pulp as an ingredient in ostrich diet: effects on meat quality. *Meat Sci* 2004;68:269–275.

[116] Wang, C, Gao, GL, Huang, JX, Zhang, KS, Zhong, H, Wang, H W, et al. Nutritive value of dry citrus pulp and its effect on performance in geese from 35 to 70 days of age. *J Appl Poult Res* 2017;26(2):253-259.

[117] Almeida, VV, Nuñez, AJC, Schinckel, AP, Alvarenga, PVA, Castelini, FR, Silva-Guillen, YV, Thomaz, MC. Interactive effect of dietary protein and dried citrus pulp levels on growth performance, small intestinal morphology, and hindgut fermentation of weanling pigs. *Anim Sci J* 2017;95(1):257-269.

[118] Crosswhite, JD, Myers, NB, Adesogan, AT, Brendemuhl, JH, Johnson, DD, Carr, CC. The effect of dietary citrus pulp on the growth, feed efficiency, carcass merit, and lean quality of finishing pigs. *Prof Anim Sci* 2013;29:345-358.

[119] Cerisuelo, A, Castello, L, Moset, V, Martinez, M, Hernandez, P, Piquer, O, et al. The inclusion of ensiled citrus pulp in diets for

growing pigs: Effects on voluntary intake, growth performance, gut microbiology and meat quality. *Livest Sci* 2010;134:180-182.

[120] Watanabe, PH, Thomaz, MC, Ruiz, US, Santos, VM, Masson, GCI, Fraga, A L, et al. Carcass characteristics and meat quality of heavy swine fed different citrus pulp levels. *Arq Bras Med Vet Zootec* 2010;62: 921-929.

[121] Brenes, A, Viveros, A, Goni, I, Centeno C, Sayago-Ayerdy, SG, Arija, I, Saura-Calixto, F. Effect of Grape Pomace Concentrate and Vitamin E on Digestibility of Polyphenols and Antioxidant Activity in Chickens. *Poult Sci* 2008;87:307-316.

[122] Goñí, I, Brenes, A, Centeno, C, Viveros, A, Saura-Calixto, F, Rebole, A, et al. Effect of dietary grape pomace and vitamin E on growth performance, nutrient digestibility, and susceptibility to meat lipid oxidation in chickens. *Poult Sci* 2007;86:508-516.

[123] Chamorro S, Viveros, A, Rebole, A, Rica, BD, Arija, I, Brenes, A. Influence of dietary enzyme addition on polyphenol utilization and meat lipid oxidation of chicks fed grape pomace. *Food Res Int* 2015;73:197-203.

[124] Aditya, S, Ohh, SJ, Ahammed, M, Lohakare, J. Supplementation of grape pomace (Vitis vinifera) in broiler diets and its effect on growth performance, apparent total tract digestibility of nutrients, blood profile, and meat quality. *Anim Nutr* 2018;4(2):210-214.

[125] Turcu, RP, Panaite, TD, Untea, AE, Șoica, C, Iuga, M, Mironeasa, S. Effects of supplementing grape pomace to broilers fed polyunsaturated fatty acids enriched diets on meat quality. *Animals* 2020;10(6):947.

[126] Gungor, E, Altop, A, Erener, G. Effect of Raw and Fermented Grape Pomace on the Growth Performance, Antioxidant Status, Intestinal Morphology, and Selected Bacterial Species in Broiler Chicks. *Animals* 2021;11(2):364.

[127] Kafantaris, I, Stagos, D, Kotsampasi, B, Hatzis, A, Kypriotakis, A, Gerasopoulos, K, et al. Grape pomace improves performance, antioxidant status, fecal microbiota and meat quality of piglets. *Animal* 2018;12(2):246-255.

[128] Yan, L, Kim, IH. Effect of dietary grape pomace fermented by Saccharomyces boulardii on the growth performance, nutrient digestibility and meat quality in finishing pigs. Asian-Aust. *J Anim Sci* 2011;24(12):1763–1770.

[129] Trombetta, F, Fruet, APB, Stefanello, FS, Fonseca, PAF, De Souza, ANM, Tonetto, CJ., et al. Effects of the dietary inclusion of linseed oil and grape pomace on weight gain, carcass characteristics, and meat quality of swine. *Int Food Res J* 2019; 26(6):1741-1746.

[130] Chedea, VS, Palade, LM, Pelmus, RS, Dragomir, C, Taranu, I. Red grape pomace rich in polyphenols diet increases the antioxidant status in key organs—kidneys, liver, and spleen of piglets. *Animals* 2019;9(4):149.

[131] Bahrami Y, Foroozandeh, A-D, Zamani, F, Modarresi, M, Eghbal-Saeid S, Chekani-Azar, S. Effect of diet with varying levels of dried grape pomace on dry matter digestibility and growth performance of male lambs. *J Anim Plant Sci* 2010;6: 605-610.

[132] Kafantaris, I, Kotsampasi, B, Christodoulou, V, Kokka, E, Kouka, P, Terzopoulou, Z, et al. Grape pomace improves antioxidant capacity and faecal microflora of lambs. *J. Anim. Physiol. Anim Nutr* 2017;101:108-121.

[133] Zhao, JX, Li, Q, Zhang, RX, Liu, WZ, Ren, YS, Zhang, CX, Zhang, JX. Effect of dietary grape pomace on growth performance, meat quality and antioxidant activity in ram lambs. *Anim Feed Sci Technol* 2018;236:76-85.

[134] Chikwanha, OC, Muchenje, V, Nolte, JE, Dugan, ME, Mapiye, C. Grape pomace (Vitis vinifera L. cv. Pinotage) supplementation in lamb diets: Effects on growth performance, carcass and meat quality. *Meat sci* 2019;147:6-12.

[135] Flores, DRM, da Fonseca, AFP, Schmitt, J, Tonetto, CJ, Junior, AGR, Hammerschmitt, RK, et al. Lambs fed with increasing levels of grape pomace silage: Effects on meat quality. *Small Rumin Res* 2021;195:106234.

[136] Tayengwa, T, Chikwanha, OC, Gouws, P, Dugan, ME, Mutsvangwa, T, Mapiye, C. Dietary citrus pulp and grape pomace as potential natural preservatives for extending beef shelf life. *Meat Sci* 2020;162:108029.

[137] Tzamaloukas, O, Neofytou, MC, Simitzis, PE. Application of Olive By-Products in Livestock with Emphasis on Small Ruminants: Implications on Rumen Function, Growth Performance, Milk and Meat Quality. *Animals* 2021;11(2):531.

[138] Tsala, A, Mpekelis, V, Karvelis, G, Tsikakis, P, Goliomytis, M, Simitzis, P. Effects of Dried Olive Pulp Dietary Supplementation on Quality Characteristics and Antioxidant Capacity of Pig Meat. *Foods* 2020;9(1):81.

[139] Ferrer, P, Calvet, S, García-Rebollar, P, de Blas, C, Jiménez-Belenguer, AI, Hernández, P, et al. Partially defatted olive cake in finishing pig diets: Implications on performance, faecal microbiota, carcass quality, slurry composition and gas emission. *Animal* 2020;14(2):426-434.

[140] Papadomichelakis, G, Pappas, AC, Tsiplakou, E, Symeon, GK, Sotirakoglou, K, Mpekelis, V, et al. Effects of dietary dried olive pulp inclusion on growth performance and meat quality of broiler chickens. *Livest Sci* 2019;221:115-122.

[141] Herrero-Encinas, J, Blanch, M, Pastor, JJ, Mereu, A, Ipharraguerre, IR, Menoyo, D. Effects of a bioactive olive pomace extract from Olea europaea on growth performance, gut function, and intestinal microbiota in broiler chickens. *Poult. Sci.* 2020;99(1):2-10.

[142] Sayehban, P, Seidavi, A, Dadashbeiki, M, Ghorbani, A, de Araújo, WAG, Durazzo, A, et al. Olive Pulp and Exogenous Enzymes Feed Supplementation Effect on the Carcass and Offal in Broilers: A Preliminary Study. *Agriculture* 2020;10(8):359.

[143] Afsari, M, Mohebbifar, A, Torki, M. Effects of Dietary Inclusion of Olive Pulp Supplemented with Probiotics on Productive Performance, Egg Quality and Blood Parameters of Laying Hens. *Annu Res Rev Biol* 2013;4(1): 198-211.

[144] Emami, A, Fathi Nasri, MH, Ganjkhanlou, M, Rashidi, L, Zali, A. Dietary pomegranate seed pulp increases conjugated-linoleic and -linolenic acids in muscle and adipose tissues of kid. *Anim Feed Sci Technol* 2015;209:79-89.

[145] Kotsampasi B, Christodoulou, V, Zotos, A, Liakopoulou-Kyriakides, M, Goulas, P, Petrotos, K, et al. Effects of dietary pomegranate byproduct silage supplementation on performance, carcass characteristics and meat quality of growing lambs. *Anim Feed Sci Technol* 2014;197:92-102.

[146] Kotsampasi, B, Christodoulou, C, Mavrommatis, A, Mitsiopoulou, C, Bampidis, VA, Christodoulou, V, et al. Effects of dietary pomegranate seed cake supplementation on performance, carcass characteristics and meat quality of growing lambs. *Anim Feed Sci Technol* 2021;273:114815.

[147] Natalello, A, Priolo, A, Valenti, B, Codini, M, Mattioli, S, Pauselli, M, et al. Dietary pomegranate by-product improves oxidative stability of lamb meat. *Meat Sci* 2020;162:108037.

[148] Ahmed, ST, Islam, MM, Bostami, RABM, Mun, H-S, Kim, Y-J, Yang, C-J. Meat composition, fatty acid profile and oxidative stability of meat from broilers supplemented with pomegranate (Punica granatum L.) by-products. *Food Chem.* 2015;188:481-488.

[149] Akuru, EA, Mpendulo, CT, Oyeagu, CE, Nantapo, CWT. Pomegranate (Punica granatum L.) peel powder meal supplementation in broilers: Effect on growth performance, digestibility, carcase and organ weights, serum and some meat antioxidant enzyme biomarkers. *Ital J Anim Sci* 2021;20(1):119-131.

[150] Gungor, E, Altop, A, Erener, G, Coskun, I. Effect of raw and fermented pomegranate pomace on performance, antioxidant activity, intestinal microbiota and morphology in broiler chickens. *Arch Anim Nutr* 2021;75(2):137-152.

[151] Tuoxunjiang, H, Yimamu, A, Li, XQ, Maimaiti, R, Wang, YL. Effect of ensiled tomato pomace on performance and antioxidant

status in the peripartum dairy cow. *J Anim Feed Sci* 2020;29(2): 105-114.

[152] Mizael, WC, Costa, RG, Rodrigo Beltrão Cruz, G, Ramos de Carvalho, FF, Ribeiro, N L, Lima, A., et al. Effect of the Use of Tomato Pomace on Feeding and Performance of Lactating Goats. *Animals* 2020;10(9):1574.

[153] Valenti, B, Luciano, G, Pauselli, M, Mattioli, S, Biondi, L, Priolo, A, et al. Dried tomato pomace supplementation to reduce lamb concentrate intake: Effects on growth performance and meat quality. *Meat Sci* 2018;145:63-70.

[154] Marcos, CN, de Evan, T, Molina-Alcaide, E, Carro, MD. Nutritive value of tomato pomace for ruminants and its influence on in vitro methane production. *Animals* 2019;9(6):343.

[155] Correia, CS, Alfaia, CM, Madeira, MS, Lopes, PA, Matos, TJS, Cunha, LF, et al. Dietary inclusion of tomato pomace improves meat oxidative stability of young pigs. *J Anim Physiol Anim Nutr* 2017;101(6):1215-1226.

[156] Biondi, L, Luciano, G, Cutello, D, Natalello, A, Mattioli, S, Priolo, A, et al. Meat quality from pigs fed tomato processing waste. *Meat Sci* 2020;159:107940.

[157] Peiretti, PG, Gai, F, Rotolo, L, Brugiapaglia, A, Gasco, L. Effects of tomato pomace supplementation on carcass characteristics and meat quality of fattening rabbits. *Meat Sci* 2013;95(2):345-351.

[158] Botsoglou, N, Papageorgiou, G, Nikolakakis, I, Florou-Paneri, P, Giannenas, I, Dotas, V, Sinapis, E. Effect of dietary dried tomato pulp on oxidative stability of Japanese quail meat. *J. Agric. Food Chem.* 2004;52:2982-2988.

[159] Sahin, N, Orhan, C, Tuzcu, M, Sahin, K, Kucuk, O. The effects of tomato powder supplementation on performance and lipid peroxidation in quail. *Poult Sci* 2008;87:276-283.

Chapter Reviewed by:

Dr. Vasiliki Kotsampasi, Associate Researcher, Research Institute of Animal Science, General Directorate of Agricultural Research, Ellinikos Georgikos Organismos-Dimitra Paralimni Giannitsa, Pella, Greece.

Dr. Michael Goliomytis, Assistant Professor of Animal Husbandry, Laboratory of Animal Breeding & Husbandry, Department of Animal Science, Agricultural University of Athens, Greece.

In: Bioactive Compounds …
Editors: G. Yildiz and G. İzli

ISBN: 978-1-68507-361-9

Chapter 5

BIOACTIVE COMPOUNDS IN PLANTS AND THEIR WOUND HEALING PROPERTIES

P. S. Sreeja Devi[1], V. S. Soumya[2] and Sreejith Parameswara Panicker[3,*]

[1]Department of Botany, Mahatma Gandhi College, University of Kerala, Kesavadasapuram, Thiruvananthapuram, Kerala, India
[2]Department of Zoology, St. Stephen's College, University of Kerala, Pathanapuram, Kollam, Kerala, India
[3]Department of Zoology, University of Kerala, Karyavattom, Thiruvananthapuram, Kerala, India

ABSTRACT

Wound healing is an intricate and active process demanding a highly regulated cascade of biochemical and cellular events designed to restore tissue integrity following injury. Studies have revealed that nature provides a vast number of plants and other products with

* Corresponding Author's Emails: psreejith@keralauniversity.ac.in; sreejadevips@gmail.com.

significant wound healing activities. Several plants have the ability to stimulate the process of wound healing. Plant extracts have different compounds that play an essential role in healing because of their natural repair mechanism. Their extracts have been found to have a profound potential in managing and treating wounds. They are reported to be safe and hypersensitive reactions rarely occur with the use of these agents. Most plants have wound healing activity due to the synergistic or additive actions of their constituents. Plant parts commonly used for wound healing are leaves followed by roots; each plant, the part's therapeutic property varies. Usually, the leaves' fresh pastes are made from the plant and are applied topically for practical use. The plant parts' medicinal property goes so that different plant parts are used in various forms, based on preparation. Modern medicine plays a significant role in treating wounds, but the number of medications is limited. Medicinal plants are frequently used as raw materials to extract active ingredients used to synthesize different drugs. Although traditional medicines offer a safe, inexpensive approach to treating wounds and burns, it has not received adequate attention. In this study, we collected the data from various Ayurvedic books and consulted with eminent doctors of Ayurveda and Modern medicine. By analyzing the parameters, selected plants can be further classified based on the type of disease to which they are used. The present study aims at disseminating the knowledge of traditional flora using discrete formulations treated for wound healing.

Keywords: ethnopharmacology, ayurveda, anti-inflammatory, traditional medicines

INTRODUCTION

A wound can be defined as the damage or disruption to the typical anatomical structure and function of the skin. Wound healing is a complex process with the wound environment, changing with the individual's changing health status. It involves coordinated interactions between diverse immunological and biological systems [1]. Various cellular and matrix components act together to re-establish damaged tissue integrity and replace lost tissue [2, 3]. Although healing is continuous, it is arbitrarily divided into different phases to help understand the physiological processes taking place in the wound and surrounding tissue [4]. Separate parts of an injury may be at different stages of healing at any one time [4-6].

As wound cause, significant deformity, and de-mobility, the search for new alternatives is under consideration. Many traditional practitioners worldwide, particularly in countries like India and China, have valuable information on many lesser-known, hitherto unknown, wild plants for treating wounds and burns [7]. Traditional forms of medicine practised for centuries in Africa and Asia are being scientifically investigated for their potential to treat wound-related disorders [8]. More than 1200 diseases are mentioned in different Ayurvedic texts. Management in various forms of these diseases is made with more than 1000 medicinal plants (89.93%), 58 minerals, metals, or ores (5.24%), and 54 animal and marine products (4.86%) [9].

The Modern Approach to Wound Healing

The wound healing drugs mainly provided by physicians are anti-microbial. Due to an increase in anti-biotic-resistant bacteria strains, wound dressings containing topical antimicrobials are not antibiotics, such as silver, iodine, honey, or polyhexamethylene biguanide (PHMB) are popular choices. The broad spectrum of these agents' anti-microbial activity may make them particularly suited to treating infection in chronic wounds, which frequently contain a wide range of microorganisms [10]. Although it is an antibiotic-silver combination, silver sulfadiazine also has broad-spectrum antimicrobial activity. It is used topically, as a cream or as part of a dressing, to treat wound infection [11]. Recent studies have demonstrated a reduction in wound infection's clinical signs while using topical antimicrobials and a statistically significant exudate decrease [12]. Systematic reviews of the use of topical antimicrobials in chronic wounds [13] have revealed a lack of robust data. This underlines the difficulties of gathering good evidence of efficacy.

Ayurvedic Approach to Wound Healing

In Ayurveda, all bodily processes are believed to be governed by a balance of the three doshas. Ayurveda has its philosophy for conceiving the physiological basis of diseases as conceived by modern western medicine. Thousands of years before, the practitioners of Ayurveda observed the physiological phenomenon of the body and patho-physiology. Specific unique approaches and procedures in Ayurveda in wound management should be emphasized here in this context. The system has described topical administration of various medications instead of instrumental procedures for incising abscess, expressing puss from suppurated ulcers, scrapping of scar and granulation tissue for accelerating wound healing etc. It also describes medications to build up fresh tissue in ulcer with severe tissue loss, to normalize the heaped surface of healed ulcers as with background tissue. The system not only aims at mere wound healing, but also describes medicinal regeneration of hair follicles and normalizing pigmentation over the wound area.

According to the Ayurveda, wounds or ulcers is the discontinuation of a lining membrane that, after healing, leaves a scar for a life closely resembling the modern definition. Viddha (punctured wound), Kshata (lacerated wound), Picchita (contusion), and Ghrista (abrasion wound) [9]. These steps have striking similarities with modern medicine. The management of wounds is described in 'Sushruta Samhita' in a very detailed manner which advised 60 methods of treating injuries.

It has been found that traditional medicine has a wide range of effective compounds as compared to modern drugs, and another advantage of conventional medicine is that there were no side effects for using plants, minerals, and animal products.

Ethnopharmacology of Wound Healing

The study of ethnomedicinal systems and herbal medicines is essential for preventing health problems in traditional communities. Since time immemorial, man has used various plants to treat and prevent many ailments [14]. There has been a significant increase in herbal formulation usage throughout the last few years in the developed world. The discovery of herbals is further complemented by knowledge on isolation, purification, characterization of active ingredients, and preparation type. The critical traditional therapeutic herbal strategy exploits several medicinal herbs to achieve extra therapeutic effectiveness, usually known as polypharmacy or polyherbalism. Specific pharmacological actions of active constituents of herbals are significant only when potentiated by other plants but not evident when used alone [15].

Other animal products used in wound healing are chameleon, in which the tail is fried in neem oil and applied to the wound site. Fat of crab, panther, tiger, wild boar, and spotted deer is used for burns, and dried earthworms are used for healing wounds. Frog skin is applied to burn injuries for rapid healing. Clear mucous produced by the garden slug is rubbed onto the skin to treat inflammations, calluses, and acne, and also it is thought to promote wound healing capacity [16]. Leech was also used for the treatment of wounds. Leech application corrects venous hypertension, reduces vascular congestion due to the presence of carboxypeptidase a inhibitors, histamine-like substances, and acetylcholine [17]; thus, it corrects venous valve dysfunction and extravascular fluid perfusion.

Considering the limitations and merits of various medicinal systems, it is inevitable to evolve an adjuvant therapy combining those systems. This is the ultimate need for the upcoming era. For instance, when we take the case of diabetic foot ulcer, which has become a comber some task of evolving an effective wound healing therapeutic strategies as far as considering the enormous statistical data and drug resistance of diabetic foot ulcer. With a diabetic prevalence of 8.6 percent, India is ranked at the top concerning the burden of diabetes [18]. Globally, the number of diabetic patients in 2011 is 366.2 million, and the projected value for 2030 is 551.8 million. Among diabetes mellitus related complications, foot ulceration is the most common, affecting approximately 15% of diabetic patients during their lifetime [19].

The number of chronic wound patients is increasing worldwide, even though the last twenty years have seen significant wound healing innovations. Successful wound management dramatically depends on the collaboration and the integration of an interdisciplinary health care team approach. Having the coordinated approach, we can improve positive outcomes through education, research, standardization and positively create a consistent delivery of evidence-based practices in wound care. Therefore, it is essential to study these plants to substantiate traditional medical knowledge and improve the treatment system.

As we know, wounds are inconvincible events in life. The prehistoric person himself has used many parts of plants and animal products for treating various skin ailments. Wounds are of different types such as cut, abrasion, incision, etc. We can classify them based on several factors. One of such classifications is based on the time of wound healing. There are two significant types of healing and non-healing wounds. Non-healing wounds pose a significant problem for the health care system worldwide, and the primary group affected by this is the elderly and diabetic population. Aging is associated in particular with increased inflammation, delayed epithelialization, and matrix degradation. Diabetic wounds are also attaining a significant increase in public concern all over the world.

Modern medicines have been widely used for treating non-healing wounds, but the fact is that there is no proper medicine designed adequately for wound healing. Many of the medications suggested by the doctors are antibiotics, painkillers, silver nitrate, glycerine, etc. In the case of antibiotics used for wound healing, the drugs provided are all the same for almost all the microbes, and other drawbacks include allergies, side effects, etc. [20].

This study can identify many of the potent plants used in Ayurvedic medicine for wound healing. This study complements the ongoing evaluation of the uses of different medicinal plants and the development of improved traditional medicine. The development of these traditional medicines with safety, efficacy, and quality will help preserve this ancestral heritage and rationalize natural products in health care. Thus, we can quickly identify rare and extinct plants for conservation and preserve the conventional practitioners' traditional heritage.

Although many of the wounds are formed in our body's external environment, the medications are given both internally, externally, or both. Washing and cleaning wounds, local application, topical administration, poultice, preparation of twine, and preparations used for expressing pus are all considered external. Internal applications of these plant preparations are relatively more minor as compared to exterior.

A plant that has been commonly used for all three and four types of wounds has been created, and plants used for wound, ulcer, and inflammatory wound were much common compared with the burn. Thirty-four plants have been used commonly for three types of diseases, and plants used against burns are significantly less widely used for wound, ulcer, and inflammatory wounds. *Diospyros malabarica* (Desr.) Kostel was the only plant used in all four types of injury, so that it is essential to study this plant's medicinal property.

The formulation prepared, by using these plant parts, was also noted during the study. The major formulations used for different types of wounds are Madhusnuhi Rasaya, Jatvadi Ghrta, Darvi Malahara, Kaisora Guggulu, Kancanara Guggulu, Tiktaka Ghrta, Saptavimsatika Guggulu, Vatari Guggulu, Pancatikta Ghrta, Yogaraja Guggulu, Draksasava and Triphala Guggulu. It was observed that *Phyllanthus emblica* is the plant species that are commonly used for many formulations. It was found that Piperaceae is the family used prominently for making all these formulations. These formulations are mainly utilized for wounds such as non-healing wounds, ulcers, diabetic carbuncle, inflammation, etc.

A detailed table of modern medicines used for wound healing was made; wounds were grouped as a cut wound, diabetic wound, burn, ulcer, and inflammatory wounds. It was observed that oral antibiotics such as Linosolidime and Cephoperazone (with Cephalosporin) were administered for cut wounds. For burns principle, antibiotics and silver sulphate were used. For diabetic wounds, commonly used medicines are Parental antibiotics and Debridise ointments. For ulcers, widely used medicines contain Mucrocin content (Mupirocin) and Glycerine and Magnesium sulphate for inflammatory wounds.

It has been observed from the study that, of the 237 plants compiled, 147 plants have been used for ulcer and 139 were being used for wounds, and a few number of plants has been used against burns. The commonly used plant part was leaves followed by root. Method of preparation is based on the plant part used. If the plant part is leaves, the leaves are pasted and applied locally over the wound area. Plants are found to be used for different types of wounds. There are 34 plants which have been commonly used for wound, ulcer, and inflammatory wound and also for burns. *Diospyrosmal abaricus* is found to be used in all four types of injuries.

The development of traditional medicinal systems incorporating plants as means of therapy can be traced back to the Middle Paleolithic Age some 60,000 years ago, as found from fossil studies. These observations suggest the importance of traditional medicine for wound healing and other etiology of the skin.

Commonly mentioned wound healing plants in the Ayurvedic books and from doctors' suggestions, and discussing with practitioners and opinion of experts were considered. Finally, a total of 237 plants have been included based on the availability of data. Some of the previous studies suggesting the plants' medicinal properties have been compiled, and they have been incorporated in the Table 1.

Anacardium occidentale, *Eupatorium odoratum*, *Euphorbia hirta*, *Ficus racemosa*, *Ixora coccinia*, *Morinda pubescens* and *Pongamia pinnata*, were studied in animal models for wound healing, analgesic, and anti-inflammatory activity based on their use in traditional medicine as wound healers, and these plants can be used to formulate drugs in pharmaceutical companies [24]. *Zingiber officinale* and *Curcuma longa*, which belong to the family Zingiberaceae, are most commonly used for their medicinal value. Traditionally, paste from the rhizomes is applied to remedy inflamed joints and sprains along with wounds and bruises [41].

Table 1. Medicinal properties of selected plants

Sl. No.	Plant name	Alignment treated	Reference
1	*Abrus precatorius* L.	Cuts and wounds	[21]
2	*Acacia catechu* Willd.	Wound	[22]
3	*Alstonia scholaris* R.Br.	Wounds	[23]
4	*Anacardium occidentale* L.	Wounds	[24]
5	*Annona squamosa* L.	Cuts and wounds	[25]
6	*Argemone mexicana* L.	Wound	[22]
7	*Aristolochia bracteolate* Lam.	Wounds	[26]
8	*Blume alacera* (Burm. f.) DC	Cuts and wounds	[27]
9	*Cassia auriculata* L.	Wound	[28]
10	*Erythrina variegata* L.	Burns, wounds	[29]
11	*Euphorbia antiquorum* L.	Burn, wounds	[24]
12	*Ficus bengalensis* L.	wounds	[24]
13	*Ficus racemosa* L.	Cuts, wounds, and ulcers	[30, 31]
14	*Holarrhena antidysentrica* (Roxb. ex. Flem.) Wall	Deep cuts, wounds	[32]
15	*Ixora coccinia* L.	Wounds	[24]
16	*Lantana camara* Linn.	Wound	[33]
17	*Morinda pubescens* J.E. Smith.	Wounds	[24]
18	*Moringa oleifera* Lam.	Cuts, wounds, snake and dogbite wounds	[34]
19	*Nerium indicum* Mill	Wound	[22]
20	*Pongamia pinnata* (L.) Pierre.	Wounds	[24]
21	*Rubia cordifolia* Linn	Ulcer	[35]
22	*Syzygium cumini* (L.) Skeels	Burns	[36]
23	*Tectona grandis*	Ulcer	[37]
24	*Tephrosia purpurea* (Linn.) Pers.	Wound	[38]
25	*Terminalia bellirica* Roxb	Wounds	[39]
26	*Terminalia chebula*	Wound	[22]
27	*Thespesia populnea* Soland	Wound	[22]
28	*Zizyphus oenoplias* (L.) Miller	Wounds	[40]

Mussaenda frondosa is used as a folk medicine to treat the wound in different parts of the world [42]. This plant has been investigated by several workers anti-microbial [43], diuretic activity [44], hepato-protective activity [45]. In Ayurveda, leaves' juice is applied on pimples, sores, ulcers & wounds to cure. The leaf extract is used for the treatment of ophthalmic disorders, erysipelas, pharygodynia, and anti-tumor. The leaves' extracts were proved to be against schward's leukemia and anti-inflammatory activities [46].

Plant products mentioned in Ayurveda need to be validated by scientific studies for their bioactive compounds. The active compounds of some of the plants for wound healing have been validated scientifically (Table 2).

Table 2. Active compounds of some of the plants for wound healing

Sl. No.	Plant name	Active compound	Reference
1	*Allium cepa* Linn.	Kampferol, alpha sitosterol, ferulic acid, myritic acid, prostaglandins	[47]
2	*Aloe vera*	Acemannan	[48]
3	*Curcuma longa* Linn.	Curumin	[49]
4	*Moringa oleifera* Linn.	Hydroproline	[50]
5	*Sesamum indicum*	Sesamin and sesamol	[51]
6	*Trigonella foenumgraceum* Linn	Folic acid	[52]
7	*Eucalyptus globulus* Labill.	Pinene, camphene, and phellandrene, citronellal, geranyl acetate	[53]
8	*Euphorbia antiquorum* L.	Euphorbin, euphol, isohelianol and camelliol C	[24]
9	*Kaemperia galanga*	Flavonoids	[54]
10	*Centella asciatica* (Linn.) Urb	asiaticosides and madicassoides	[55]

Allium cepa belonging to the family Liliaceae contains kaempferol, sitosterol, ferulic acid, myritic acid and prostaglandins. These constituents used as an abortifacient, bulb extract shown to have ecobolic effect in rats. Also, the plants were proved to have anti-diabetic, anti-oxidant, anti-hypertensive, anti-thrombotic, hypoglycemic, antihyperlipidemic properties [56]. Acemannan is the major polysaccharide of Aloe vera, which stimulates VEGF expression and other wound healing-related factors. *Azadiracta indica* leaf extracts and oil from seeds have a proven anti-microbial effect. In *Aegel marmelos*, the fruit pulp is red with a gummy and astringent taste. The chief constituent of the drug is marmelosin which is furocoumarin. The drug also contains carbohydrates, protein, volatile oil, and tannins. The pulp also contains vitamin C and vitamin A. two alkaloids Omethylhalfordional and isopentylhalfordinol have been isolated from fruits. It is used as a digestive appetizer and also used in the treatment of diarrhea and dysentery. It is also a tonic, and it has wound healing properties [57].

Wound infection is one of the most common diseases in developing countries because of poor hygienic conditions [58]. Ayurveda has comprehensive wound care management [59]. The plant-based materials are used as first aid antiseptic coagulants and wound wash [60]. Recently, developed countries have turned to traditional medicinal systems involving herbal drugs and remedies [61]. It is often noted that 25% of all medications prescribed today come from plants [62]. This estimate suggests that plant-derived medicines make up a significant segment of natural product-based pharmaceuticals. Many Ayurvedic plants have a vital role in the process of wound healing. Most plants' wound healing property is due to the synergistic or additive actions of their constituents. Plant-based therapy accelerates the healing process and maintains aesthetics [63]. More than 13,000 plants have been studied during the last five years [64]. The concept of an alternative treatment system, notably herbal and Ayurvedic medicines therapy, is gaining ground and attracting attention worldwide.

Many of the studies are done on medicinal plants. The plants were taken based on the geographical distribution, availability of the medicinal plants, and information from the local practitioners in the area under study. Our study has been based on the details of various medicinal plants referred to in different ayurvedic books and based on the parameters of whether they are used for wounds. A significant number of plants have been used for ulcers and injuries, but only a few plants were used against burns. It was found that many animal products have been especially aiming at the treatment of burns than wounds. The fat obtained from many animals has been used for treating burns [65]. The primary group of animals and animal products used for the burn has been detailed in various studies.

All the plant parts are highly nutritious and also found to be associated with therapeutic value, but the major plant part used for treating wounds are leaves. It was observed from previous studies that leaves are generally very cleansing, healing, soothing, and revitalizing as well as nourishing [66]. As leaves are the site of photosynthesis, they are an essential part of the plant body. Leaves of many plants contain secondary metabolites stored in them. Leaves were the preferable plant part used and were mainly applied topically as a paste, powder, or sap on the affected area [67]. Other plant parts such as flowers and fruits were used less because they are available only in their flowering and fruiting season. A paste can be easily made using leaves, and it will be more effective when we use new medications than others. Studies have proven that many plants' leaves contain many vitamins such as niacin, pyridoxine, pantothenic acid, riboflavin, and minerals such as copper, potassium, calcium, manganese, iron, selenium, zinc, and magnesium.

The plant products are usually pasted and are typically administered on the wound. Wounds are mainly formed on the skin's surface so that the plant products are made into a paste and applied topically or sometimes made into a poultice. The topical application of poultice helps to reduce inflammation and edema. Many plant bark and roots contain medicinal properties, and decoction has been made from these plants, usually used to wash wounds to reduce infection. Traditionally, medicinal plants have been used for many years as topical and internal preparations to promote wound repair [68]. The medicines are usually pasted or ground because they can provide all the phytochemical products present in the plant. The product will be more effective; in any other form, such as decoction, the plant's volatile substance may be lost. Different parts of the same plant are also used for the treatment. The study has found that both oral and external administration of wound healing recipes can promote the restoration of chronic refractory ulcers. The effect of combined administration is superior to a single administration. The oral, exterior, and integrated application can improve wound healing in the inflammatory phase [69].

Thirty-four plants have been noted, commonly used for all three types of wounds, but plants used for burn were not found to be widely used for other wounds. Relatively plants for the burn are more miniature. *Diospyros malabarica* is the only plant that has been found to be used for all four types of diseases. The plant's methanol extract seemed to have an antioxidant property [58]. Many other studies were also done on this plant [70]. But few studies were done for validating the medicinal property of this plant; due to that, further studies should be done for validation. The plant species which is commonly used in many of the significant formulations is *Phyllanthus emblica*. There have been many detailed studies done on the medicinal property of this plant. They were seemed not only as wound healing agents but also found to be anti-diabetic, anti-bacterial, promoting hair growth, reducing cholesterol, and having anti-cancerous properties [71]. Numerous experimental pieces of evidence have shown that amla fruit possesses antioxidant [72], hepatoprotective [73], hypocholesterolemic [74], and anti-inflammatory activities [75].

Drug formulation in *Ayurveda* is based on two principles: Use as a single drug and use of more than one drug, of which the latter is known as Poly Herbal Formulation. This key traditional therapeutic herbal strategy exploits several medicinal herbs to achieve extra therapeutic effectiveness, usually known as polypharmacy or polyherbalism [76]. Even though individual plants' active phytochemical constituents have been well established, they are typically present in minute amounts, and always, they are insufficient to achieve the desired therapeutic effects. For this, scientific studies have revealed that these plants of varying potency, when combined, can theoretically produce a more significant result, as compared to individual use of the plant and also the sum of their effect. There are various formulations used for wound healing, and the synergy of plant combinations can have an even more significant impact on treating wounds than single drugs [77]. These combinations are used externally to treat gangrene of wounds [78]. Due to synergism, polyherbalism confers some benefits not available in a single herbal formulation. It is evident that a better therapeutic effect can be reached with a single multi-constituent formulation. In formulations, there will be combined action of plants and animal products such as ghee, and from various studies, it was found that the plants' synergistic action makes the wounds heal quickly. Different combinations were used, ranging from two plants in a variety to more plants were combined [77].

The plant family which is commonly used in many of the formulations was Piperaceae. The genus Piper, belonging to the Piperaceae family, possesses about 700 species distributed in tropical and subtropical regions [79]. Some of these species are economically important such as *Piper nigrum* L. (black pepper), *Piper methysticum* (kava), and *Piper betle* (betel leaf), and others are used as condiments and as medicinal [80]. Piper leaves contain distinctively aromatic and acrid volatile oils that contain such compounds and elements as cadinene, carvacrol, caryophyllene, chavibetol, chavicol, eugenol, terpinyl, acetate [81], piperine, piperlongumine, pyridine alkaloids, sesamin, tannins, oxalic acid and iron [82, 83].

Modern medicines were widely used for treating wounds, but the fact is that many of these medicines do not entirely aim at the process of wound healing. *Antimicrobials*, including iodine-based preparations and silver releasing agents, are used to treat infected wounds (there may be a dose-dependent effect). Antimicrobial agents target bacteria at several levels (cell membrane, cytoplasmic organelle, and nucleic acid), thus minimizing bacterial resistance. Antimicrobial agents such as silver, povidone-iodine, and poly- hexa- methylene- bi- guanideare sometimes incorporated into dressings to control or prevent infection [84].

Other plant products used for wound healing include resin, heartwood, rhizome bulb, etc. Heartwood and resin of many plants contain phytochemicals; several previous studies suggest that some trees provide these parts as the medicinal agent. Usually, decoctions were prepared from this plant part. Antimicrobial properties were observed in the resin of many plants during phytochemical analysis [85]. Heartwood mainly contains many essential oils, and these oils were found to be anti-inflammatory. This property of the plant can be applied in treating wounds [86]. Essential oils present in these plants were found to act against the bacterial infections that affect them, so essential oil in the plant part was considered the second line of defense.

Scientists, doctors, and traditional healers are now working together to improve conventional knowledge for the effective use of herbal remedies. Over the past 20 years, our understanding of the wound healing process has increased drastically. The development of new, exciting technologies that accelerate normal wound healing and counter the pathophysiologic processes that lead to chronic wound formation has come along with this knowledge. From growth factors to bioengineered skin substitutes, the future of wound healing holds great promise.

FINAL REMARKS

Pharmacological validation of Indian medicinal plants and animal products used in wound healing is minimal. A large number of these ethnopharmacological products with enormous potential have not been validated for their wound healing properties. With the advent of synthetic drugs, plant drugs lost their significance for some time. However, ethnomedicine has gained considerable importance in the recent past, because of being safe and having no side effects. A total of 237 plant species widely used for wound healing has been consolidated in our study. Out of the 237 plants, 147 has been used to treat ulcer, 139 for wound, 96 plants for inflammatory wound, and 16 plants were used for the burn. The number of plants used for the burn is comparatively less. The plant parts mainly used for wound healing are leaves, followed by roots. Many of the plant parts are selected based on their availability.

According to the difference in the plant part used in wound healing, the preparation method also varies. A commonly used form is a paste that is usually freshly prepared and applied topically over the wound area. Besides, it was found that plant combinations are both externally and internally administered for an effective result. The plant's commonly used family was also noted, and it was found that Fabaceae were, the most widely used family for wound healing. The leguminous plants are found to be very useful in a nutritious and also in a medicinal way. For the preparation of various formulations, the commonly used family is Piperaceae which were found to have several condimental and therapeutic properties. The leaves of the plants contain several aromatic and acrid volatile acids.

It was found from the study that *Diospyros malabarica* and *Phyllanthus emblica* seem to have many medicinal properties. *Diospyros malabarica* was found to have been used for all four different types of wounds. The methanol extract of the plant was proved to have anti-oxidant properties. *Phyllanthus emblica* is a plant commonly used for the preparation of many formulations. This plant was found to have wound healing properties and anti-diabetic, anti-bacterial, promoting hair growth, reducing cholesterol, and anti-cancerous properties.

Many of the medications were also used in the treatment of wounds. They were oral antibiotics, principal antibiotics, silver sulphate, glycerin, magnesium sulfate, and many antibiotics. But the major drawback of this modern medicine is that side effects such as allergies, and these medications do not target the exact process of wound healing.

The study revealed that nature provides many plants and other products with significant wound healing properties. These natural agencies are a rich target for the development of alternatives to synthetic drugs. The combination of traditional and modern knowledge can produce better medications for wound healing with lesser side effects. However, there is a need for scientific validation, standardization, and safety evaluation of plants, animal products, and formulations of traditional medicine before these could be recommended to heal wounds.

Acknowledgments

We thank Kerala State Plan funds, Kerala Biotechnology Commission, YIPB program, KSCSTE; HRD scheme, Dept. of Health Research-Start up Grant, Govt. of India, and the University of Kerala.

References

[1] Glat P M, Longaker M T. Wound healing. In: *Grabb and Smith's Plastic Surgery.* 5th edn. Lippincott–Raven, Philadelphia 1997; pp. 3–12.

[2] Rothe M, Falanga V. Growth factors, their biology and promise in dermatologic disease and tissue repair. *Arch. Dermatology* 1989; 125:1390.

[3] Shakespeare P. Burn wound healing and skin substitutes. *Burns* 2001; 27:517.

[4] Richardson M. Acute wounds: an overview of the physiological healing process. *Nurs Times* 2004; 100:50.

[5] Hart J. Inflammation. 2: Its role in the healing of chronic wounds. *J. Wound Care* 2002; 11:245.

[6] Broughton G, Janis J E, Attinger C E. The basic science of wound healing. *Plast. Reconstr. Surg*. 2006; 117: 12.

[7] Kumar B, Vijayakumar M, Govindarajan R, Pushpangadan P. Ethnopharmacological approaches to wound healing – Exploring medicinal plants of India. *J. Ethnopharmacol*. 2007; 114:103.

[8] Krishnan P. The scientific study of herbal wound healing therapies. Current state of play: *Curr. Anaes. Crit. Care* 2006; 17: 21–27.

[9] Biswas T K, Mukarjee B. Plant medicine as Indian origin for wound healing activity. *International journal of lower extremity wounds* 2003; 2:25.

[10] Davies K G. Interactions between nematodes and microorganisms: bridging ecological and molecular approaches. *Ad. Appl. Microbiol.* 2005; 57:53-78.

[11] White R J, Cooper R. Silver sulphadiazine: a review of the evidence. *Wounds UK* 2005; 1: 51-6.

[12] Kotz P, Fisher J, McCluskey P, Hartwell S D, Dharma H. Use of a new silver barrier dressing, Allevyn Ag in exuding chronic wounds. *Int. Wound J.* 2009; 6 (3): 186-94.

[13] Vermeulen H, van Hattem J M, Storm-Versloot M N, Ubbink D T. Topical silver for treating infected wounds. *Cochrane Database Syst. Rev.* 2007; 24.

[14] Chah K F, Eze C A, Emuelosi C E, Esimone C O. Antibacterial and wound healing properties of methanolic extracts of some Nigerian medicinal plants. *J. Ethnopharmacol.* 2006; 104:164.

[15] Subramani P, Thing G S and Dhanaraj SA. Polyherbal formulation: Concept of Ayurveda. *Pharmacogn. Rev.* 2014; 8:73.

[16] Quave C L, Andrea Pieroni, Bradley C Bennett. Dermatological remedies in the traditional pharmacopoeia of Vulture. *Journal of Ethnobiology and Ethnomedicine* 2008; 4:5.

[17] Andreas M, Manfred R, Gustav D. *Medicinal Leech Therapy*. Gustav Dobos, New York 2007; pp 11-12, 132-138.

[18] Whiting D R, Guariguata L, Weil C, Shaw J. I D F Diabetes Atlas: Global estimates of the prevalence of diabetes for 2011 and 2030. *Diabetes Res. Clin. Pract.* 2011; 94:311.

[19] Shankhdhar K L K, Shankhdhar U, Shankhdhar S. Diabetic foot problems in India: An overview and potential simple approaches in a developing country. *Current Diabetes Reports* 2008; 8:452.

[20] Michael J K, Cynthia A, Kirr R N. Wound Care, *Wild Iris Medical Education*, Inc. All Rights Reserved 2012.

[21] Katewa SS, Chaudhary BL, Anita Jain. Folk herbal medicines from the tribal area of Rajasthan, India. *Journal of Ethnopharmacology* 2004;92(1):41-46. DOI:10.1016/j.jep.2004.01.011.

[22] Patil S B, Naikwade N S, Kondawar M S, Magdum C S, Awalel V B. Traditional uses of plants for wound healing in the Sangli

district, Maharashtra. *International Journal of Pharm. Tech. Research* 2009; 1:876.

[23] Bhardwaj S, Gakhar S K. Ethnomedicinal plants used by the tribals of Mizoram to cure cuts and wounds. *Indian Journal of traditional Knowledge* 2005; 4:75.

[24] Ayyanar M, Ignacimuthu S. Herbal medicines for wound healing among tribal people in Southern India: Ethnobotanical and Scientific evidences. *International Journal of Applied Research in Natural Products* 2009; 2:29.

[25] Nayak S, Dan D, Julien R Marshall, Godwin N Isitor. Wound-healing Properties of the Oils of *Vitis vinifera* and *Vaccinium macrocarpon*. *Phytotherapy Research* 2011;25(8):1201-8. doi:10. 1002/ptr.3363.

[26] Ramadas A V, Gunesh S, Thomas G A O, Williams GT, Hawthorne AB. Natural history of Crohn's disease in a population-based cohort from Cardiff (1986-2003): a study of changes in medical treatment and surgical resection rates. *Gut* 2010;59:1200–6.

[27] Rajendran R, Hasobo Mohammad Ahamed, Radhai R, Balakumar C. In Vitro Evaluation of Wound Healing Efficiency of Herbal Nanoparticles Loaded Anti-microbial Wound Dressing. *Proceedings of the 4th International Conference on Nanostructures (ICNS4)* 12-14 March, 2012, Kish Island, I.R. Iran.

[28] Chopda M Z, Mahajan R T. Wound Healing Plants of Jalgaon District of Maharashtra State, India. *Ethnobotanical Leaflets* 2009; 13:1.

[29] De Britto, John, Mahesh R. Exploration of Kani Tribal Botanical Knowledge in Agasthiayamalai Biosphere Reserve - South India. *Ethnobotanical leaflets* 2007; 11(18): 258-265.

[30] Punjani B L. Ethnobotanical aspects of some plants of Aravali Hills in North Gujarat. *Ancient Sc. of Life* 2002; 21: 268– 80.

[31] Sen S K, Behara L M. Ethnomedicinal plants used against skin diseases in Bargat district of Orissa. *Ethnobot* 2003; 15: 90– 96.

[32] Parinatha M, Harish G U, Vivek N C, Mahesh T, Shivanna M B. Ethno-botanical wealth of Bhadra Wildlife Sanctuary in Karnataka. *Indian Journal of Traditional Knowledge* 2004; 3: 37–50.

[33] Kurian J C. Plants that heal. Owners Oriental Watchman Publishing House, *Pune* 1995; 190.

[34] Begum D, Nath S C. Ethnobotanical review of medicinal plants used for skin diseases and related problems in North-eastern India. *J. Herbs Spices Med. Plants* 2000; 7: 55.

[35] More B H, Gadgoli C, Padesi G. Hepatoprotective activity of *Rubia cordifolia*. *Pharmacology online* 2007; 3:73.

[36] Mishra D K, Samanta C, Mishra T K. Ethnomedicine of the Tribe Khasia of Midnapore District, West Bengal. *Journal of Economic and Taxonomic Botany Additional Series* 1996; 12: 82–91.

[37] Majumdar M, Nayeem N, Kamath J, Mohammed A. Evaluation of *Tectona grandis* leaves for wound healing activity. *Pak. J. Pharm. Sci.* 2007; 20: 120-4.

[38] Chaudhari T B, Dipak A, Tambe, Sanjay R C, Machindra J C, Dinesh PH. An Investigation of Wound Healing Activity of *Tephrosia purpurea* (Linn.) Pers. Root extracts in Rats. *Journal of Pharmacy Research* 2010; 3:2215.

[39] Choudhary G P. Wound healing activity of the ethanol extract of *Terminalia bellirica* Roxb. Fruits. *Natural Product Radiance* 2008; 7:19.

[40] Maya S, Kumari S, Menon P S V. Ethnobotanical notes on the flora of sacred tanks of Kerala. *Ethnobot.* 2003; 15:55-59.

[41] Eigner D, Scholz D. *Ferula asa-foetida* and *Curcuma longa* in traditional medical treatment and diet in Nepal. *J. Ethnopharmacol.* 1999; 67:1

[42] Kapil A I, Koul O P, Suri. Anti-hepatotoxic effects of chlorogenic acid from *Anthocephalus cadamba*. *Phytother Res* 1995; 9:189.

[43] Basu B D, Kirtikar K. *Indian medicinal plants.* 1st edn. New Delhi 1998; pp 1481.

[44] Patil S A, Joshi V G, Sutar P S. Screening of alcoholic extract of mussaenda from dosa leaf for wound healing and antibacterial activities in albino rats. *Pharmacologyonline* 2010; 2:761.

[45] Sambrkar S N, Patil P A, Kangralkar V A. Protective activity of *Musssaenda frondasa* leaf extract against paracetamol induced hepatic damage in wistar rats. *Journal of Pharmacology Research* 2010; 3.

[46] Chandayak. The wealth of India, *A dictionary of Indian raw materials industrial products New Delhi.* Council of Scientific and Industrial Research 1991; pp 28 -30.

[47] Shenoy C, Patil M B, Kumar R, Patil S. Preliminary phytochemical investigation and wound healing activity of *Allium cepa* Linn (Liliaceae): *International Journal of Pharmacy and Pharmaceutical Sciences* 2009; 2:167.

[48] Jettanacheawchankit S, Sasithanasate S, Sangvanich P, Banlunara W, Thunyakitpisal P. Acemannan stimulates gingival fibroblast proliferation; expressions of keratinocyte growth factor-1, vascular endothelial growth factor, and wound healing. *J. Pharmacol. Sci.* 2009; 109:525.

[49] Rao S G V, Selvaraj J, Senthil R, Radhakrishnan R N, Murali, Manokar B. Efficacy of some indigenous medicines in wound healing in rats. *Indian Journal of Animal Sciences* 2003; 73:652.

[50] Rathi B S, Bodhankar S L, Baheti A M. Evaluation of aqueous leaves extract of *Moringa oleifera* Linn. for wound healing in albino rats. *Indian Journal of Experimental Biology* 2006; 44:898.

[51] Rajeswari S V, Thiruvengadam N M, Ramaswamy. Production of interspecific hybrids between *Sesamum alatum* Thonn and *Sesamum indicum* L. through ovule culture and screening for phyllody disease resistant. *S. Afr. J. Bot.* 2010; 76:252.

[52] Sumitra M, Manikandan P, Suguna L, Hittar G C E. Study of Dermal Wound Healing Activity of *Trigonella foenumgraceum* Seeds in Rats. *J. Clin. Biochem. Nutr.* 2000; 28:59.

[53] Hukkeri VT, Karadi RV, Akki KS, Savadi RV, Jaiprakash B, Kuppast J, Patil MB. Wound healing property of *Eucalyptus globulus* leaf extract. *Indian Drugs* 2002; 39:481.

[54] Azarkan M, Wintjens R, Looze Y, Baeyen VD. Detection of three wound induced proteins in papaya latex. *Phytochemistry* 2004; 65:225.

[55] Shetty B S, Udupa S L, Udupa A L, Somayaji S N. Effect of *Centella asiatica* L. (Umbelliferae) on normal and dexamethasone suppressed wound healing in Wistar Albino rats. *J. Lower Extremity Wounds* 2006; 5:137.

[56] Avasthi B K, Tewari J D. A Preliminary Phytochemical Investigation of *Desmodium gangeticum* DC. (Leguminosae). *J. Am. Pharm. Assoc. Am. Pharm. Assoc. (Baltim)* 1995; 44:625.

[57] Jaswanth A, Loganathan V, Manimaran S, Rukmani S. Wound healing activity of *Aegle marmelos*. *Indian J. Pharma. Sci.* 2001; 63:41.

[58] Kumar S M, Sripriya R, Vijaya R H, Sehgal P. Wound Healing Potential of *Cassia fistula* on Infected Albino Rat Model. *J. Surg. Res.* 2006; 131:283.

[59] Vinaya P N, Prasad J S R A. Moist wound healing: perspectives from Ayurveda. *Int. Res. J. Pharm.* 2013; 4.

[60] Kumar B, Vijayakumar M, Govindarajan R, Pushpangadan P. Ethnopharmacological approaches to wound healing – Exploring medicinal plants of India. *J. Ethnopharmacol.* 2007; 114:103.

[61] Lanfranco G. Popular Use of Medicinal Plants in the Maltese Islands. *Insula* 1992; 1:34.

[62] Raskin I, Ripoll C. Can an apple a day keep the doctor away. *Curr. Pharm. Des.* 2004; 10: 3419.

[63] Galal E E, Gawad M A. Anti-diabetic activity of Egyptian onion *Allium cepa* extract. *J. Egypt Med. Assoc.* 1965; 48:14.

[64] Biswas T K, Mukarjee B. Plant medicine as Indian origin for wound healing activity. *International journal of lower extremity wounds* 2003; 2:25.

[65] Padmanabhan P, Sujana K A. Animal products in traditional medicine from Attapadi Hills of Wesrtern Ghats, Division of Forest Ecology and Biodiversity Conservation, Kerala Forest Research Institute, Peechi 680620, Kerala. *Indian Journal of Traditional Knowledge* 2008; 7:326.

[66] Iwu M. *Handbook of African Medical Plants.* (CRC Press. Boca Raton Fl.1993) 1993; Pp. 21 – 27.

[67] De W H, SibongileNciki, Sandy F van Vuuren. Medicinal plants used for the treatment of various skin disorders by a rural community in northern Maputaland, South Africa. *Journal of Ethnobiology and Ethnomedicine* 2013; 9:51.

[68] Odimegwu D C, Ibezim E C C, Esimone C O, Nworuand C S, Okoye F B C. Wound Healing and Antibacterial Activities of the Extract of *Dissotis theifolia* (Melasto-Mataceae) Stem Formulated in a Simple Ointment Base. *Journal of Medicinal Plant Research* 2008; 2:011.

[69] Shen L, Que HF. Effects of oral and external application of recipes for promoting granulation in rats with chronic skin ulcers. *Zhong Xi Yi Jie He Xue Bao* 2009; 7:1150.

[70] Amrish C, Tarasingh R R. *Pharmacognostic evaluation of Diospyros malabarica bark powder*. Amity Institute of Pharmacy, Noida 2012; 2:178.

[71] Singh E, Sharma S, Pareek A, Dwivedi J, Yadav S, Sharma S. Phytochemistry, traditional uses and cancer chemopreventive activity of Amla *(Phyllanthus emblica*). *The Sustainer: Journal of Applied Pharmaceutical Science* 2011; 02: 176.

[72] Chaudhari R K. Emblica cascading antioxidant: A novel natural skin care ingredient: *Skin pharmacology and applied skin physiology* 2002; 15:374.

[73] Jeena K J, Joy K L, Kuttan R. Effect of *Phyllanthus emblica*, *Phyllanthus amarus* and *Picrorhiza kurroaon* N-Nitrosodiethylamine induced hepatocarcinogenesis. *Cancer Letters* 1999; 136:11.

[74] Mishra M, Pathak U N, Khan A B. *Emblica officinalis* Gaertn and serum cholesterol level in experimental rabbits. *British Journal of Experimental Pathology* 1981; 62:526.

[75] Asmawi M Z, Kankaanranta H, Moilanen E, Vapaataslo H. Anti-inflammatory activities of *Emblica officinalis* Gaertn leaf extracts. *Journal of Pharmacy and Pharmacology* 1993; 45:581.

[76] Subramani P, Thing G S, Dhanaraj S A. Polyherbal formulation. Concept of ayurveda: *Pharmacogn. Rev.* 2014; 8:73.

[77] Bhat P, Hegde G, Hegde G R. Ethnomedicinal practices in different communities of Uttara Kannada district of Karnataka for treatment of wounds. *J. Ethnopharmacol.* 2012; 143:501.

[78] Van Wyk B E, Gericke N. *People's Plants.* Pretoria, Briza Publications 2000.

[79] Guimaraes E F, Giordano L C S. Piperaceae do Nordestebrasileiro I estado do Ceara [Piperaceae from the Northeast of Brazil I state of Ceará]. *Rodriguésia* 2004; 55:21.

[80] Joy B, Sandhya CP, Remitha KR. Comparison and bioevaluation of *Piper longum* fruit extracts. *J. Chem. Pharm. Res.* 2010; 2:696.

[81] Dyer L A, Richards J, Dodson C D. *A model genus for studies of evolution, chemical ecology, and trophic interactions.* Kluwer Academic Publishers, Boston 2004; pp117-139.

[82] Waard D, Anunciado I S. *Piper nigrum* L. in Plant Resources of South-East Asia No. 13. *Spices.* Backhuys Publishers, Leiden 1999.

[83] Teo S P, Banka R A. *Piper betle* L. in Plant Resources of South-East Asia No. 16. *Stimulant Plants.* Backhuys Publishers, Leiden 2000.

[84] Motta G J, Milne C T, Corbett L Q. Impact of antimicrobial gauze on bacterial colonies in wounds that require packing. *Ostomy Wound Manag.* 2004; 50:48.

[85] Arno S. *Coniferous resin salve, ancient and effective treatment for chronic wounds* – laboratory and clinical studies Department of Orthopedics and Traumatology, Helsinki University Hospital University of Helsinki, Finland 2013.

[86] Tumen I, Süntar I, Eller F J, Keleş H, Akkol E K. Topical wound-healing effects and phytochemical composition of heartwood essential oils of *Juniperus virginiana* L., *Juniperus occidentalis* Hook, and *Juniperus ashei*. J Buchholz, *J. Med. Food.* 2013; 16:48.

Chapter Reviewed by:

Dr. *Ajaikumar B. Kunnumakkara*, PhD., Professor, Cancer Biology Laboratory; Coordinator, DAICENTER (DBT-AIST International Center for Translational and Environmental Research) at IITG, Department of Biosciences and Bioengineering, Indian Institute of Technology Guwahati, Guwahati, Assam-781039, India.

Dr. *P. Kanagaraj,* MSc., M.Phil., PhD., Assistant Professor (S.G), DST-SERB Project Principal Investigator, Cancer Science Laboratory, Department of Biotechnology, School of Bioengineering SRM Institute of Science and Technology, Kattankulathur, Chennai, – 603203, India.

In: Bioactive Compounds …
Editors: G. Yildiz and G. İzli
ISBN: 978-1-68507-361-9

Chapter 6

Biodiversity and Health Effects of Sesquiterpenes in Aromatic and Medicinal Plants

Martina Jakovljević Kovač[1]***, Maja Molnar***[1,*]
and Igor Jerković[2]
[1]Josip Juraj Strossmayer University of Osijek,
Faculty of Food Technology Osijek, Osijek, Croatia
[2]University of Split, Faculty of Chemistry and Technology,
Split, Croatia

Abstract

The importance of natural products in the treatment of various health conditions, as well as in drug discovery, is impressive. This is supported by the fact that over three quarters of the world's population mainly relies on plants and herbal products for health care. According to the study provided by World Health Organization in 2002, more than

[*]Corresponding Author's Email: mmolnar@ptfos.hr.

30% of plant species in the world were used for medicinal purposes. The presence of various bioactive compounds is the main reason for the widespread use of aromatic and medicinal herbs, both in folk and modern medicine. These bioactive components, showing positive health effects in various health conditions, are the focus of numerous studies.

Terpenes, a group of highly bioactive plant components, play an important role in aromatic plants considering the ecological chemistry of interaction with insects and pathogens. Terpenes are one of the main constituents of the plant essential oils, most commonly used in the flavor and fragrance industry. Their presence in the plants is highly dependent on the plant species and growing conditions. Many essential oil terpenes possess significant positive health effects and pharmaceutical importance, which encourages further research on their mode of formation, characterization and biological activity.

Terpenes are classified according to their skeletal structure, with the most common being monoterpenes, diterpenes and sesquiterpenes listed as a group of 15-carbon isoprenoid compounds derived from the assembly of three isoprenoid units. In higher plants, sesquiterpenes are represented by several acyclic, mono, bi-, tri- and tetracyclic systems, many of which possess significant biological activity such as antimicrobial, antitumor, and cytotoxic properties. Thousands of sesquiterpenes known today are formed from 300 different C15-hydrocarbon skeletons that are produced from the substrate farnesyl diphosphate (FPP) by the action of sesquiterpene synthases.

This chapter highlights the biodiversity of plant sesquiterpenoids as well as their health effects, focusing mainly on modern health problems. The appearance of different types of sesquiterpenes as well as a comparison of their structure and biological activity, with the emphasis on anticancer activity, is reviewed.

Keywords: biodiversity, biological activity, cancer, essential oil, sesquiterpenes

INTRODUCTION

Given that over three-quarters of the world's population relies on plants and herbal products to maintain their health, it is no surprise that many approved drugs originated from natural sources [1]. There is still a great interest in the use of medicinal plants for health purposes, especially in the prevention of inflammatory responses. The use of natural anti-inflammatory products can provide a safer and more acceptable alternative to many common widely-used synthetic drugs [2]. Bioactive components defined by Biesalski et al. [3] as "*either essential or nonessential compounds found in nature or part of the food chain and showing a particular effect on human health,*" among others, include the components such as terpenes, phenolic acids, flavonoids, glycosides and alkaloids.

Terpenes as highly bioactive plant components are present in aromatic and medicinal plants, especially in their essential oils. Due to their numerous positive health effects, the plants and plant essential oils have become useful in many health conditions. Their structural diversity greatly depends on the biosynthetic pathways employed in their formation, which then contributes to the wide variety of biological activities justifying their application in the food and pharmaceutical industry [4]. Despite their numerous biological activities, the biosynthesis of only a few sesquiterpenes *in vivo* or in cell-free enzyme systems has been investigated [5].

Biodiversity of Sesquiterpenes

Terpenes are the largest and structurally most diverse group of natural compounds [6]. To date, more than 30,000 natural terpenes have been identified, most of which are present in the plants where they play an important role in the ecological chemistry of interaction with insects and pathogens. The essential oils are rich in terpenes, and their type and concentration in the plants depend on many factors, such as the plant species, growing conditions, etc. and they are often used in the flavor and fragrance industry [1].

Despite a wide variety of terpenes, they are all formed in the biosynthetic pathway involving the fusion of C5 units with an isopentenoid structure. They are divided according to the skeletal structure with the emphasis on the number of isoprenoid units. Therefore, 15-carbon compounds, sesquiterpenoids are composed of three isoprenoid units that can exist as acyclic, mono-, bi-, tri-, and tetracyclic systems [1]. The biosynthetic pathway of terpene formation consists of four phases, the first of which involves the formation of isopentenyl diphoshate (IPP), synthesized in the plants in two ways. IPP, along with dimethlyallyl diphosfate (DMAPP), as an allylic isomer, can be formed using mevalonic acid pathway, or methylerythritol phosphate (MEP) pathway. The second phase involves condensing the basic C5 units to form larger prenyl diphosphates, geranyl diphosphate (GPP, C10), farnesyl diphosphate (FPP, C15) and geranylgeranyl diphosphate (GGPP, C20). In the third stage, C10–C20 diphosphates undergo numerous cyclizations and rearrangements to form the parent carbon skeletons for each group of terpenes. In the third phase, the products of the second phase are converted to terpenes by converting GPP to monoterpenes, FPP to sesquiterpenes and GGPP is converted to the diterpenes. In addition, FPP and GGPP have the ability to dimerize to form precursors of the C30 and the C40 terpenoids. In the last stage, various transformations occur, such as oxidation, reduction, isomerization and conjugations, and the basic structure of terpenes is converted into thousands of different derivatives [7].

As already mentioned, in the first step of terpene biosynthesis, C5 building blocks are formed using the mevalonate or the methylerythritol phosphate pathway. The mevalonate pathway involves the condensation of three molecules of acetyl coenzyme A (AcCoA) to form C6 structure, also referred as 3-hydroxy-3-methylglutaryl-CoA (HMG-CoA). This is followed by the conversion of HMG-CoA to mevalonic acid catalyzed by HMG-CoA reductase, which requires the presence of nicotinamide adenine diphosphate (reduced form) (NADPH) to carry out the reaction. For the formation of the C5 molecule, IPP, from mevalonic acid, two phosphorylations and decarboxylation elimination is required [7].

As for the second biosynthetic pathway, a non-mevalonate pathway, the most significant progress occurred in 1994 when Cartayrade et al. [8] investigated mechanistic pathways for the isoprenoid biosynthesis in *Ginkgo biloba*. After that, according to the research on *E. coli* mutants, it has been shown that the precursors of this new pathway are glyceraldehyde phosphate and pyruvate. However, the first step of this pathway was modified by assuming that hydroxyethylthiamine diphosphate, the C2 unit formed from pyruvate, condenses with glyceraldehyde-3-phosphate to form 1-deoxy-D-xylulose-5-phosphate catalysed by deoxyxylulose phosphate synthase [9] (Rohmer et al., 1996). In this pathway, the reactions including deoxy-D-xylulose-5-phosphate reductoisomerase are involved in the transformation of 1-deoxy-D-xylulose-5-phosphate into the isoprenoid carbon skeleton, 2-*C*-methyl-D erythritol-4-phosphate. This transformation involves the cleavage of the C3–C4 bond of deoxyxylulose backbone as well as the formation of a new link between C3 and C4. In order for IPP to be formed from 2-C-methyl-D-erythritol-4-phosphate, several reactions, including dehydration steps, reductions and phosphorylation are required [7]. According to Vranová et al. [10,11], (*E*)-4-hydroxy-3-methylbut-2-enyl diphosphate is formed from 2-C-methyl-D-erythritol-4-phosphate in the following steps, by the action of several enzymes, such as 2-C-methyl-D-erythritol-4-phosphate cytidylyltransferase, CDP-ME kinase, 2-C-methyl-D-erythritol-2,4-cyclodiphosphate synthase and 1-hydroxy-2-methyl-2-(*E*)-butenyl-4-diphosphate synthase. In the last step, IPP and DMAPP are produced from (*E*)-4-hydroxy-3-methylbut-2-enyl diphosphate by the action of (*E*)-4-hydroxy-3-methylbut-2-enyl diphosphate reductase. In this pathway, IPP and DMAPP are formed in a ratio of 5 or 6 to 1.

In the plant cells, terpenes are produced in plastids and cytosol [12, 13]. Generally, monoterpenes and diterpenes are formed in the plastids, while sesquiterpenes are produced in the cytosol/ER (endoplasmic reticulum) compartment. Furthermore, there is a difference in the pathway involved in terpene production, whereas the initial step of terpene biosynthesis in the plastids takes place by methylerythritol phosphate pathway, while in the cytosol/ER mostly by mevalonate pathway [7]. However, it appears that both pathways need to be included for the biosynthesis of certain terpenes, as in the case of sesquiterpenes of chamomile (*Matricaria recutita*). Hereby, the first two C5 units are formed *via* methylerythritol phosphate pathway and the third unit is formed by mevalonate pathway and glyceraldehyde – pyruvate pathway [14].

Formation of C5 units is followed by the joining of the basic C5 building blocks, IPP and more reactive allylic isomer, DMAPP, to form larger intermediates such as C10, C15 and C20 prenyl diphosphates. DMAPP is formed in the reversibile reaction from IPP by the action of the enzyme IPP isomerase. The process of adding IPP units to DMAPP in sequential chain elongation steps is catalysed by the enzyme prenyltransferases. The head-to-tail condensation (1`-4) of one IPP and DMAPP produces the C10 molecule, called geranyl diphospate (GPP), and further condensation of IPP with the resulting allylic diphosphate chain leads to the formation of the C15 unit, farnesyl diphosphate (FPP). In the next step, IPP can bind to the resulting C15 chain to form C20 molecule, known as geranylgeranyl diphosphate (GGPP). The prenyltransferases that catalyze the formation of GPP, FPP, and GGPP are important regulatory enzymes in the plant terpene biosynthesis, since they, as well as their localization, directly affect the emergence of various major classes of terpenoids [7].

The next step involves the action of the enzyme terpene synthases, also known as terpene cyclases, since most of the products formed are in the cyclic form. A formation of cyclic products utilizes a carbocationic reaction mechanism that is very similar to that used by prenyltransferases. It has been shown that the reaction usually begins with the cleavage of the metal ion-assisted diphosphate group [15-17] whereas an enzyme-bound allylic cation is formed, which further undergoes isomerization, i.e., electrophilic attack on the central or distal double bond, whereby a cyclization occurs to yield new cations [1, 18].

Naturally occurring sesquiterpenes, can be divided into seven groups based on their parent cation as well as the first step of cyclization that occurs in the process of biosynthesis. The cyclization reactions can be divided into two types, one type involving cyclization of the initially formed farnesyl cation to form 10-membered (*E*,*E*)-hermacradienyl cation or 11-membered (*E*,*E*)-humulyl cation rings. Since these are large rings, (*E*)-geometry of the C2–C3 double bond is not a limitation during cyclization. In the second cyclization type, which occurs after the isomerization of the C2–C3 double bond, tertiary nerolidyl cation is formed, and its cisoid conformer undergoes cyclization to form the central or distal double bond, forming different cations, such as 1,6-(bisabolyl cation), 1,7-(cycloheptanyl cation), 1,10-((*Z,E*)-hermacradienyl cation) or 1,11-((*Z,E*)-humulyl cation). Sesquiterpene synthase mechanisms, in addition to the cyclization, include 1,2-, 1,3- and 1,4- hydride shifts, ring contractions, Wagner–Meerwein rearrangements and 1,2-methyl shifts, and they contribute to such a wide range of different sesquiterpenes [19]. Due to subsequent rearrangements, a large number of different structures can be formed [1, 18].

All sesquiterpenes known today are formed from 300 distinct C15-hydrocarbon skeletons, which are derived from the single substrate, FPP in the cytosol, using sesquiterpene synthases. Given that there is a large number of different terpene synthases as well as that some synthases can produce multiple products, terpene diversity can be attributed to the terpene synthases [19]. Nevertheless, terpene skeletal diversity is associated not only with the number of terpene synthases but also with the ability of these catalysts to form multiple products from a single substrate. More than half of known terpene synthases, in addition to the main products, lead to the formation of additional products, which was first discovered during the sequential purification steps of the plant extracts [20]. The same has been demonstrated with deuterium labeled substrates, where it has been shown that a decrease in the production of one product leads to an increase in the production of another, which suggests that the components arise from a common intermediate of a single enzyme [21, 22]. Furthermore, the same has been proven with terpene synthase genes, where it has been shown that the application of one synthase produces a larger number of products and can be attributed to the fact that various reactive carbocationic intermediates can be stabilized in a number of ways. On the other hand, since half of sesquiterpene synthases carbocationic based catalysis produces only one final product, it can be concluded that the properties of the enzyme rather than the mechanism, are responsible for the formation of one or more products [19]. In addition, the formation of more products can also be stimulated by more conformational flexibility in the active center allowing the formation of more reaction intermediates, which in turn leads to the formation of more products. According to the structural elucidation of complexes, intermediates present in thermodynamically preferred conformations have shown that terpene synthase reactions are more likely to be controlled by kinetic rather than thermodynamic processes [23]. Therefore, the formation of more products is also affected by the difference in substrate conformation that occurs at very beginning of the reaction.

Sesquiterpenes are divided into several classes such as acyclic, mono-, bi-, tri-, and tetracyclic systems as well as lactones [24].

Biosynthesis of Acyclic Sesquiterpenes

The reaction mechanism of acyclic sesquiterpene formation begins with the formation of FPP. Subsequently, metal-mediated (Mg^{2+} ionization) removal of the diphosphate anion in the FPP substrate leads to the formation of farnesyl cation. Farnesyl cation can be isomerized to the cisoid (2*Z*,6*E*)-farnesyl cation (nerolidyl cation). Sesquiterpene alcohols, (*E*, *E*)-farnesol, (*Z*,*E*)-farnesol, (3*R*,6*E*)-nerolidol and (3*S*,6*E*)-nerolidol are formed from these cations by the addition of water. The protonation of farnesyl cation results in (*E*)-α-farnesene and (*E*)-β-farnesene [25]. According to Cordell [26], farnesol biosynthesis is one of the most investigated fields in the biosynthesis of natural compounds. Farnesol is produced without oxidation at C-5 of mevalonate, whereby the stereochemical integrity of the carbon of mevalonate is maintained. However, upon conversion to farnesol, each unit of mevalonate loses hydrogen atom from C-4.

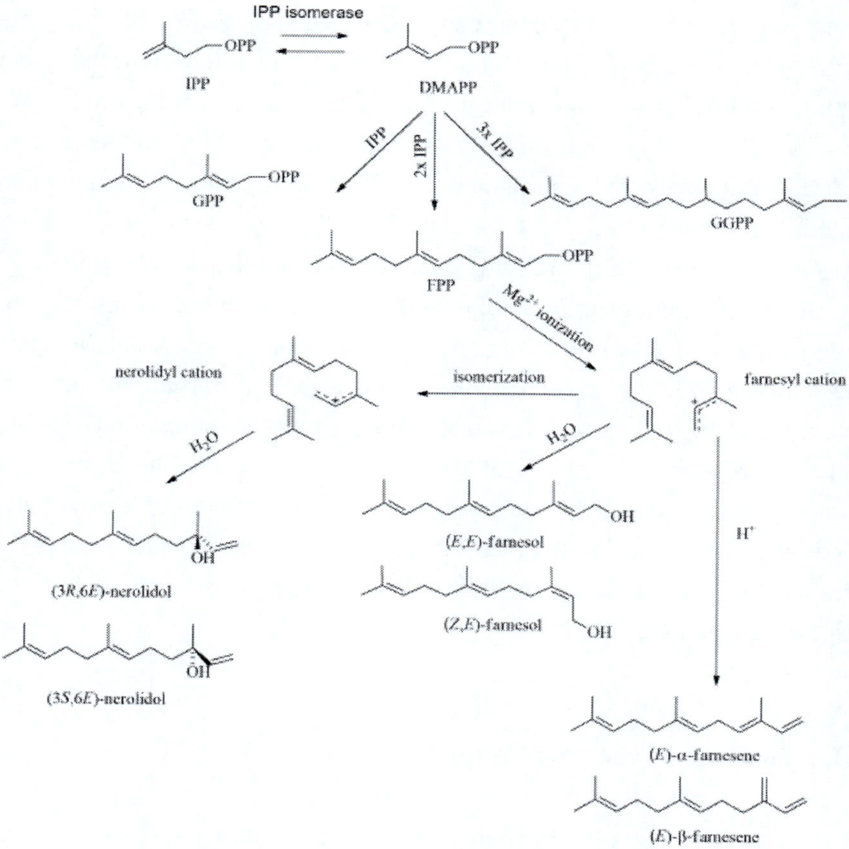

Figure 1. A proposed pathway for the biosynthesis of the most common acyclic sesquiterpenes [7, 25].

Biosynthesis of Monocyclic Sesquiterpenes

The reaction mechanism of cyclic sesquiterpene formation begins in the same way as the acyclic synthesis with the formation of farnesyl and nerolidyl cation that undergo different cyclizations. Farnesyl cation can undergo either *via* 10-*exo-trig* (attack a double bond center whereby, with respect to the emerging ring in the transition state, the pair of electrons is placed in the displaced bond, or exo to the developing ring) or 11-*endo-trig* (attack a double bond center with a pair of electrons forming part of the newly formed ring) cyclizations on the C10-C11 double bond to form (*E*,*E*)-germacradienyl cation and (*E*,*E*)-humulyl cation. On the other hand, nerolidyl cation can undergo four different cyclizations, *via* 10-*exo-trig* or 11-*endo-trig* where C1 attacks the C10-C11 double bond to form (*Z*, *E*)-germacradienyl cation and (*Z*,*E*)-humulyl cation. On the other hand, the cyclization to the C6-C7 double bond *via* 6-*exo-trig* or 7-*endo trig* can occur, resulting in bisaboyl cation and cycloheptanyl cation [25].

The possible biosynthesis of α-humulene in the plants begins with the cyclization of FPP to (*E*, *E*)-humulyl carbocation which is catalyzed by sesquiterpene cyclase. After the cyclization, α-humulene, as 11-membered ring compound, is formed from humulyl carbocation [27, 28].

α-Bisabolene, a pure hydrocarbon molecule, is formed in the catalytic process of ring closure from bisabolyl cation intermediate, where water does not intervene. On the other hand, in the case where water attacks the carbocation intermediate, α-bisabolol is formed, as sesquiterpene with alcohol group [29, 30].

Biosynthesis of Bicyclic Sesquiterpenes

Characterization of caryophyllane sesquiterpenes is possible due to the specific bicycle [7.2.0] undecane ring system, called the caryophyllane skeleton, where dimethylcyclobutane and nine-membered ring containing *trans*-endocyclic (4–5) double bond join. This basic structure is subjected to different rearrangements and cyclization reactions, where different derivatives can be formed [28]. In plants, the possible biosynthesis of (*E*)-β-caryophyllene and its derivatives begins with the cyclization of FPP to (*E*,*E*)-humulyl carbocation, catalyzed by sesquiterpene cyclase. Humulyl carbocation is then converted into caryophyllenyl cation, which has bicyclic structure containing 4- and 9-membered rings. The formation of isocaryophyllene derivatives requires anticlockwise rotation of the initial form, i.e., the enzymatic *E-Z* isomerization. The second derivative, β-caryophyllene oxide, is formed by few oxidation steps of β-caryophyllene [27, 28, 31]. In the case of δcadinene, the bicyclic sesquiterpene with naphthalene-type ring skeleton, nerolidyl cation is required, whose 1,10-cyclization results in (*Z*, *E*)-germacradienyl cation, followed by 1,6-closure to give ring partitioning [19].

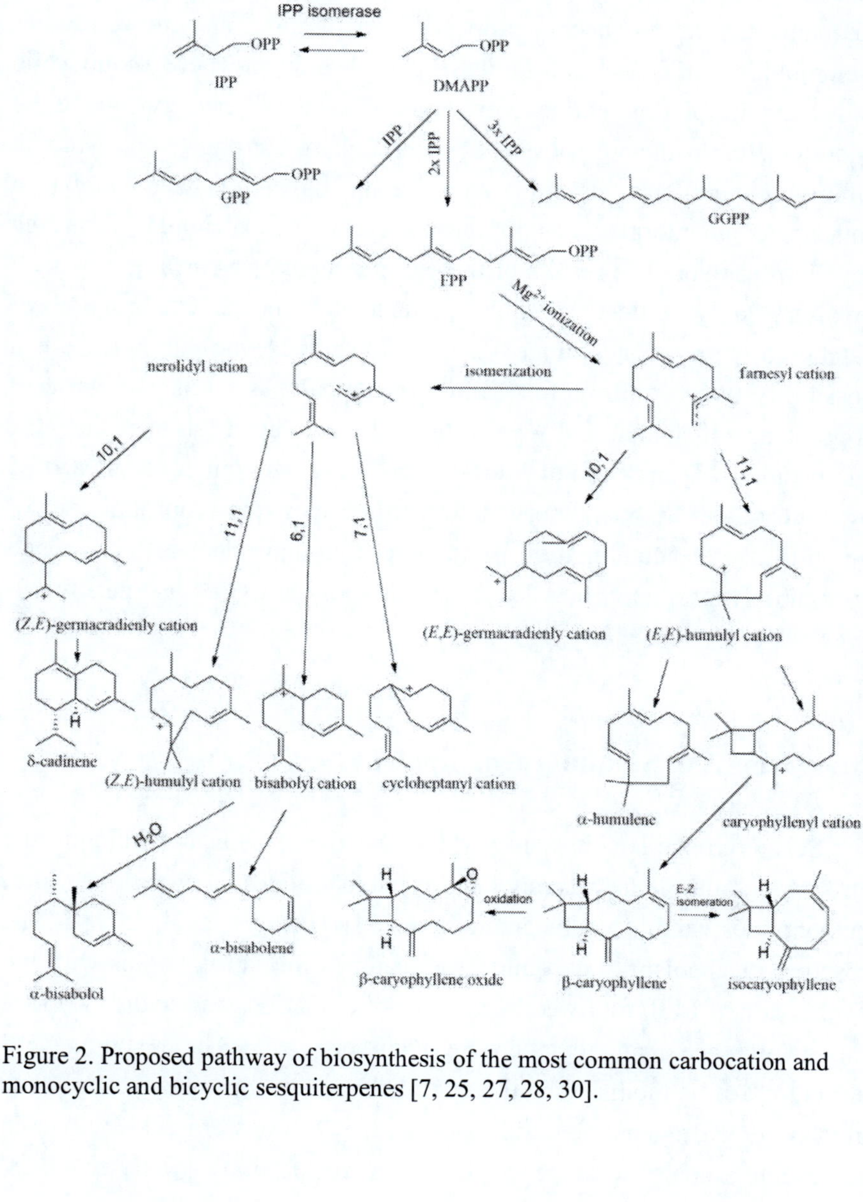

Figure 2. Proposed pathway of biosynthesis of the most common carbocation and monocyclic and bicyclic sesquiterpenes [7, 25, 27, 28, 30].

Biosynthesis of Tricyclic Sesquiterpenes

Patchouli alcohol (patchoulol), tricyclic sesquiterpene, is one of the main components of the essential oil of *Pogostemon cablin*, and due to its unique woody fragrance is a common ingredient in perfumes. Several mechanisms have been proposed involving enzyme-catalyzed, electrophilic cyclization of farnesyl pyrophosphate where monocyclic and bicyclic olefins germacrene and bulnesene were considered as intermediates in the formation of patchoulol. According to the isotopic dilution experiments, sesquiterpene olefins have not been shown to participate as intermediates in the formation of patchoulol from the acyclic precursor. The results of these experiments support a mechanism involving only tertiary cationic intermediates in the construction of patchoulol with a concomitant 1,3-hydride shift. Nevertheless, there is a possibility that certain olefins such as patchoulenes might still arise as coproducts of patchoulol cyclization. It has been shown that the cyclization of patchoulol does not have to include nerolidyl pyrophosphate as mandatory intermediate but the coproduction of nerolidol by the combination of farnesyl pyrophosphate isomerase and pyrophosphatase activities leads to the possibility of isomerization-cyclization sequence [32].

Biosynthesis of Sesquiterpene Lactones

Sesquiterpene lactones are divided into five groups, according to the structure, into germacranolides, elemanolides, eudesmanolides, guaianolides, and pseudoguaianolides [33] (Babaei et al. 2018). The reaction mechanism of sesquiterpene lactones formation begins with the cyclization of FPP, followed by a series of oxidations and hydroxylations that take place using cytochrome P450 enzymes. The resulting C15 structure can be modified by the application of various enzymes, such as alcohol dehydrogenases, reductases and acyl transferases [34] which represents a complex sequence of enzymatic reactions strictly regulated in each step. The biosynthesis, accumulation and expression of genes required for the biosynthesis of sesquiterpene lactones are also influenced by a number of external factors such as light, stress and the presence of pathogens, as well as with internal factors, such as type of tissue and stage of plant development [10].

The biosynthetic route for artemisinin formation is the first sesquiterpene lactone route that is fully characterized. The first reaction involving the formation of amorpha-4,11-diene from FPP was catalyzed by the enzyme amorpha-4,11-diene synthase. This is followed by hydroxylation reaction of amorpha-4,11-diene involving cytochrome P450 monooxygenase and cytochrome P450 oxidoreductase, to artemisinic alcohol. The resulting alcohol is then oxidized to the aldehyde using alcohol dehydrogenase and then converted to dihydroartemisinic aldehyde. The penultimate step involves the formation of dihydroartemisinic acid from dihydroartemisinic aldehyde by an aldehyde dehydrogenase, which is an immediate precursor of artemisinin. This process is considered as non-enzymatic photooxidation, although this part is still under discussion. In addition to the above, there is an alternative route by which artemisinic aldehyde can be transformed to artemisinic acid by the action of aldehyde dehydrogenase [35].

In the formation of sesquiterpenes lactones, as in other sesquiterpenes, an important role is played by germacrene A synthase, which converts FPP to germacrene A, the precursor of many germacranolide-type sesquiterpene lactones. Germacrene A is a backbone skeleton for the formation of costunolides and parthenolides whose importance is manifested in their anticancer activity. In a process involving several oxidation reactions, germacrene A is converted to germacra-1(10),4,11(13)-trien-12-oic acid (GAA) using germacrene A oxidase (cytochrome P450-like enzyme). The resulting acid is further oxidized to costunolide using costunolide synthase, where epoxidation of the C4 – C5 double bond occurs to form the final parthenolide product [35].

HEALTH BENEFITS OF SESQUITERPENES

Sesquiterpenes are known for their biological and pharmaceutical activities, such as antimicrobial, antitumor, anti-inflammatory and cytotoxic properties. In the plants, they arise as the plant defense mechanisms from insects and microbes and act as attractants, deterrents, antifeedants and phytoalexins [1]. In the last two decades, sesquiterpenes have shown positive health effects in a number of metabolic conditions such as hyperglycemia, hyperlipidemia, cardiovascular diseases, neural difficulties, diabetes and cancer [36].

Anti-Inflammatory Activity

Inflammation is a physiological and pathological process in which a complex cascade of different molecular and cellular signals participates. Inflammation in the body is an adaptive response whose role is to protect and treat infection or injury [37]. Numerous studies have shown how essential oils and their constituents, with an emphasis on sesquiterpenes, represent a promising source of anti-inflammatory substances [38].

Farnesol (3,7,11-trimethyl-2,6,10-dodecatriene-1-ol), as the acyclic sesquiterpene alcohol present in the essential oils of many plants including lemon grass (*Cymbopogon citratus*) and rose, has been proven to possess anti-inflammatory effects [39]. The studies in mice have shown the action of farnesol in reducing inflammation, by regulating the expression of type 1 T helper (Th1) cytokine IFN-γ (interferon gamma) and Th1-inducing cytokine IL-12 (interleukin 12) [40]. Furthermore, farnesol has also been shown to play a protective role in lung inflammation and reduce the oxidative stress caused by cigarette smoking as it stimulates endogenous defense enzymes in rat lungs. The components found in cigarette smoke form reactive oxygen species (ROS) which can lead to DNA damage and interfere with the work of vital organs such as the lungs. Malondialdehyde (MDA) is one of the substances derived from ROS and is considered a biomarker for oxidative stress and an indicator of lipid peroxidase [40]. Furthermore, the substances present in cigarette smoke lead to an increase in epithelial permeability and inflammatory reactions, as well as various lung diseases. Administration of farnesol (50 and 100 mg / kg), once a day for one week to rats that intratracheally received cigarette smoke extract, led to a decrease in total cell count, lactate dehydrogenase levels, as well as the proportion of protein and malondialdehyde (MDA) in BALF (bronchoalveolar lavage fluid) [41. Farnesol, not only reduced the effects of exposure to the components, but increased the proportion of glutathione as the cellular antioxidant, decreased the proportion of hydrogen peroxide by restoring catalase levels as the enzyme that breaks down hydrogen peroxide into water and oxygen, and restored the levels of redox cycle components [42]. The anti-inflammatory effect and reduction of oxidative stress are also seen in colon cells of Wistar rats. As in the lung, oral supplementation with farnesol (50 and 100 mg/kg) over seven days showed a significant reduction in oxidative stress by reducing the tissue lipid peroxidation process. Also, an increase in the levels of redox cycle components has been observed with an increase in the activity of superoxide dismutase, catalase and quinone reductase. The level of these enzymes is very important given their role in the capturing free radicals while protecting the colonic tissue [43]. Besides, farnesol reduced the activity of caspase-3, the enzyme involved in the cleavage of cellular proteins and DNA fragmentation during the apoptosis process [44].

Bisabolol (6-methyl-2-(4-methyl-3-cyclohexen-1-yl)-5-hepten-2-ol), also known as α-bisabolol, as monocyclic sesquiterpene has been found as an important compound in the essential oil German chamomile (*Matricaria recutita*) and *Myoporum crassifolium*. Due to several positive health properties with an emphasis on skin benefits, bisabolol is very often used in the cosmetic industry [45]. The anti-inflammatory activity of bisabolol is manifested in the ability of bisabolol to interfere with ROS production thereby improving the antioxidant network and redox balance [46].

β-Caryophyllene [4,11,11-trimethyl-8-methylene-bicyclo [7.2.0] undec-4-ene], as a bicyclic sesquiterpene has been found in many plants, as an important compound in the extracts and essential oil, and is found in the essential oil of cloves (*Syzygium aromaticum*) [47], hemp (*Cannabis sativa*) [48], and rosemary (*Rosmarinus officinalis*) [49]. It has been shown that β-caryophyllene affects the endocannabinoid system, more specifically binding to one of the receptors in this pathway, CB2 (cannabinoid receptor type 2) and thus reducing the ability of the receptor to mediate cellular signals that could ultimately lead to inflammation. In addition, β-caryophyllene acts on other signaling pathways, thus inhibiting the pathways leading to the formation of proinflammatory cytokines, and on the other hand improves the action of Th1 immunity [7, 50]. Using different approaches, it has also been shown that β-caryophyllene (0.1–100 μg/mL) can act as the chemopreventive agent in protecting cultured human lymphocytes in pre- and co- treatment from damage caused by ethyl methanesulfonate and colcemid, while in post-treatment caryophyllene does not show protective effect [51]. It can often be found in combination with the monocyclic sesquiterpene, α-humulene, and its application is widespread in the fragrance and flavoring area. The studies of anti-inflammatory activity of β-caryophyllene and α-humulene show that several different signaling pathways are involved in this process [48]. Ethanol extract of *Cordia verbenacea* D.C. (Boraginaceae), the brazilian medicinal plant used in many health care products, contains a number of components, the most common of which are α-humulene and β-caryophyllene, which at doses of 50 mg/kg orally have anti-inflammatory effects. The essential oil of this plant shows a number of anti-inflammatory activities and effects on immunity by reducing tumor necrosis factor (TNF-α), without affecting interleukin 1 beta [52]. The anti-inflammatory activity of these two components was also tested on a model of acute inflammation in Wistar rat paw, which among other things, includes neutrophil recruitment, cytokine production, activation of MAPKs (mitogen-activated protein kinase) and NF-κB (nuclear factor-κB), and upregulated expression of quinine B1 receptors. α-Humulene and β-caryophyllene have been shown to reduce neutrophil migration and

NF-κB activation, while α-humulene decreased TNF-α and IL-1β levels. None of the components had an inhibitory effect on the activation of MAPKs (ERK (extracellular-signal-regulated kinase), p38 (p38 mitogen-activated protein kinases), and JNK (c-Jun N-terminal kinase) and did not antagonize directly on B1 receptors, which are considered a target for anti-inflammatory drugs since they affect inflammation and pain in inflammatory processes. Nevertheless, α-humulene exhibits dexamethasone-like anti-inflammatory activity, decreasing B1 receptor mRNA levels [52-54]. The anti-inflammatory activity of these two components has also been proven experimentally in the case of allergic airways. If we look at α-humulene, in mice it was shown to reduce inflammatory activity by reducing inflammatory mediators, adhesion molecule expression as well as activation of transcription factors [55].

In addition to β-caryophyllene, caryophyllene oxide is also one of the substances often used in the food, pharmaceutical and cosmetic industries. This sesquiterpene derivative was first identified in the extract of *Annona squamosa* [56]. Caryophyllene α-oxide has been shown to have an anti-inflammatory effect on rat and mice paw edema at doses less than the commonly used doses of aspirin [57].

Chamazulene (7-ethyl-1,4dimethylazulene), as bicyclic sesquiterpene that is formed from the sesquiterpene matricin is present in many plants such as wormwood (*Artemisia absinthium*), yarrow (*Achillea millefolium*), and chamomile (*Matricaria chamomilla*), and occurs during hydrodistillation from the guajane-type sesquiterpenelactone matricin [14]. Chamazulene, in contrast to matricine, has been shown to be beneficial for the antiinflammatory activity and to prevent the peroxidation of arachidonic acid in chamomile extract in a concentration-dependent manner [58].

Patchouli alcohol as a tricyclic sesquiterpene, has attracted a great attention due to its potential in cognitive enhancing, learning impairment attenuating, and as a neuroprotective and anti-inflammatory agent. According to Xian et al. [59] the anti-inflammatory activity of patchoulol can be attributed to the down-regulation of mRNA expression of few inflammatory mediators, for example TNF-α, IL-1β, IL-6, iNOS (nitric oxide synthase), and COX-2 (prostaglandin-endoperoxide synthase 2).

Anti-Cancer Activity

Currently, cancer is one of the most serious diseases, which is the second most common cause of death, after heart disease. Cancer occurs due to various abnormalities in normal growth and homeostasis. If we look at the cellular level, these abnormalities include increased proliferation, evasion of apoptosis, angiogenesis, and the possibility of metastasis to parts distant from the primary site. In tumor development, it is important to mention the interaction between incipient tumor cells and other cells, as well as the entire microenvironment in which they are located. Chemotherapy is the most commonly used treatment, which involves the use of cytotoxic drugs on cancer cells. However, there are many problems and limitations with the use of chemotherapy with an emphasis on their adverse side effects and low efficacy [33]. Despite numerous studies and advances in cancer research, there is still a need for new, effective cancer drugs with as few harmful side effects as possible. The application of natural products with an emphasis on bioactive components, as one of the oldest methods, is still being investigated. This category includes terpenes as well, which have being investigated as potential anticancer agents due to their biodiversity [1].

Acyclic sesquiterpene alcohol farnesol was shown to be an effective initiator of cell cycle arrest and apoptosis in various cancer cells. Furthermore, in various animal models it was shown that through the exposure to several different cell cycles, it can inhibit tumor genesis in the initiation or progression stage, being considered a chemopreventative and anti-tumor agent *in vivo*. The main mechanism of action of farnesol which inhibits cell proliferation and stimulates apoptosis is considered to be the stimulation of endoplasmic reticulum stress and activation of the unfolded protein response. Also, farnesol is thought to participate in other cycles, such as nuclear receptor-signaling pathways including PPARα (peroxisome proliferator-activated receptor) signaling pathway and NF-κB pathway [39].

The importance of α-bisabolol, monocyclic sesquiterpene possessing a 1,4-functionalized cyclohexane skeleton and present in the essential oils of various plants, is manifested in the cytotoxic effect on human and rat glioma cells. α-Bisabolol was shown to has a strong time- and dose-dependent cytotoxic effect, and a dose of 10 mM is required for the death of 100% glioma cells apoptotic induction [1].

Numerous previous studies on caryophyllane sesquiterpenes have shown that they can block carcinogen-mediated DNA damage, show antiproliferative and chemosensitizing activities in cancer cells, and protect of noncancerous cells from anticancer drug toxicity. In particular, a number of effects on different signaling pathways have been demonstrated for β-caryophyllene, with emphasis on NF-κB, STAT3 (signal transducer and activator of transcription 3), PI3K/Akt (phosphoinositide 3-kinase/ protein kinase B), TNF-α, and redox cascades, such as Nrf2 (nuclear erythroid 2-related factor 2) and MAPK, likely through a CB2R-activation. Also, β-caryophyllene, β-caryophyllene oxide, and α-humulene act synergistically as chemo-therapeutic drugs which is associated with ABC-transporter inhibition, whereby they can enable low-dose chemotherapeutic treatments and affect chemoresistance occurrence. If we compare caryophyllane sesquiterpenes, α-humulene and isocaryophyllene, caryophyllane sesquiterpenes show a higher cytotoxicity power which is associated with *cis*-configuration of caryophyllane scaffold and the exocyclic double bond [28].

Sesquiterpene lactones, particularly parthenolide, allantolactone, costunolide and dehydrocostus lactone, isoalantolactone, deoxyelephantopin and costunolide, show an anticancer effects through changes in the redox cell balance and effects on signaling pathways, especially in the case of NF-kB and STAT3 pathway. Their influence is reflected through the reduction of factors involved in the cell cycle, the stimulation of apoptosis of cancerous cells through growth in apoptotic factors and the reduction of anti-apoptotic factors [33].

Table 1. Overview of the activities of well-known sesquiterpenes as well as the corresponding plant sources

The most common plant sources	Determined sesquiterpenes	Type of sesquiterpenes	Determined activity	Reference
Santalum spicatum, Cymbopogon nardus, Cymbopogon citratus	Farnesol (2*E*,6*E*)-3,7,11-trimethyldodeca-2,6,10-trien-1-ol	Acyclic	Anticandida Antiinflammatory Chemopreventative Anticarcinogenic	[39, 41, 43, 60]
Melaleuca quinquenervia, Virola surinamensis	Nerolidol (6*E*)-3,7,11-trimethyldodeca-1,6,10-trien-3-ol	Acyclic	Antiinsecticidal Antimalarial	[61, 62]
Matricaria recutita, Myoporum crassifolium	Bisabolol (2*R*)-6-methyl-2-[(1*R*)-4-methylcyclohex-3-en-1-yl]hept-5-en-2-ol	Monocyclic	Antiirritant, Antiinflammatory Antimicrobial Anticarcinogenic	[1, 45, 55
Salvia officinalis, Mentha spicata, Lindera strychnifolia Uyaku	α-Humulene (1*E*,4*E*,8*E*)-2,6,6,9-tetramethyl-1,4,8-cycloundecatriene	Monocyclic	Antiinflammatory Antibacterial Antifungal Antiproliferative	[28, 52, 63, 64]
Cordia verbenacea, Syzygium aromaticum, Cannabis sativa	β-Caryophellene (1*R*,4*E*,9*S*)-4,11,11-trimethyl-8-methylidenebicyclo[7.2.0]undec-4-ene	Bicyclic	Antiinflammatory Anticarcinogenic Analgesic Antiallergic Antiarthritic Antibacterial Antifungal Antiproliferative	[28, 52, 63, 64, 66]

The most common plant sources	Determined sesquiterpenes	Type of sesquiterpenes	Determined activity	Reference
Annona squamosal, Centaurea calcitrapa, Centaurea choulettiana, Centaurea pullata, Cannabis sativa	Caryophyllene oxide (1*R*,4*R*,6*S*,10*R*)-4,12,12-trimethyl-9-methylidene-5-oxatricyclo[8.2.0.0^{4,6}]dodecane	Bicyclic	Antiinflammatory Antiallergic Analgesic Antibacterial Antifungal Antiproliferative	[28, 57, 66]
Matricaria chamomilla, Artemisia absinthium, Achillea millefolium	Chamazulene (7-ethyl-1,4-dimethylazulene)	Bicyclic	Antiinflammatory Antioxidant Antitumor	[58, 67]
Patchouli cablin	Patchoulol (1*R*,4*S*,4a*S*,6*R*,8a*S*)-4,8a,9,9-Tetramethyldecahydro-1,6-methanonaphthalen-1-ol)	Tricyclic	Antiviral against influenza Antiinflammatory	[59, 68, 69]
Artemisia annua	Artemisinin (1*S*,4*S*,5*R*,8*S*,9*R*,12*S*,13*R*)-1,5,9-trimethyl-11,14,15,16-tetraoxatetracyclo [10.3.1.04,13.08,13]hexadecan-10-one)	Lactone	Anticarcinogenic Antioxidant Antimalarial	[70, 71]
Tanacetum parthenium	Parthenolide (1*S*,2*S*,4*R*,11*S*)-4,8-dimethyl-12-methylidene-3,14-dioxatricyclo[9.3.0.02,4]tetradec-7-en-13-one)	Lactone	Anticarcinogenic Antioxidant	[72]
Radix inulae, Inula japonica, Rudbeckia Subtomentosa, Inula royleana	Alantolactone (3a*R*,5*S*,8a*R*,9a*R*)-5,8a-Dimethyl-3-methylene-3a,5,6,7,8,8a,9,9a-octahydronaphtho[2,3-*b*]furan-2(3*H*)-one)	Lactone	Anticarcinogenic Antioxidant	[73, 74]

Table 1. (Continued)

The most common plant sources	Determined sesquiterpenes	Type of sesquiterpenes	Determined activity	Reference
Saussurea lappa, Vladimiria souliei, Aucklandia lappa, Laurus nobilis	Costunolide (3a*S*,6*E*,10*E*,11a*R*)-6,10-dimethyl-3-methylidene-3a,4,5,8,9,11a-hexahydrocyclodeca[b]furan-2-one)	Lactone	Anticarcinogenic Antioxidant	[75, 76, 77]
Radix inulae, Inula japonica, Rudbeckia Subtomentosa, Inula royleana	Isoalantolactone (3a*R*,4a*S*,8a*R*,9a*R*)-8a-methyl-3,5-dimethylidene-3a,4,4a,6,7,8,9,9a-octahydrobenzo[f][1] benzofuran-2-one)	Lactone	Anticarcinogenic Antioxidant	[78]
Elephantopus scaber, Elephantopus carolinianus	Deoxyelephantopin [(3*S*,4*R*,8*R*,9*E*,12*R*)-10-methyl-5-methylidene-6,14-dioxo-7,13-dioxatricyclo[10.2.1.04,8] pentadeca-1(15),9-dien-3-yl] 2-methylprop-2-enoate	Lactone	Anticarcinogenic Antioxidant	[79, 80]
Valeriana officinalis	Valerenic acid (pentanoic acid)	Lactone	Antiinflammatory Sedative Anxiolytic	[4]

Given that numerous scientific studies have shown anticancer potentials of sesquiterpenes and their derivatives through the inhibition of cell proliferation, induction of apoptosis in tumor cells and angiogenesis, several sesquiterpene lactones are currently in clinical trials (Phase I and II). Tested sesquiterpene lactones were shown to be multi-functional as they act *via* different signaling pathways [36]. According to many researches, the nucleus of sesquiterpene lactones is considered important and essential for the development of new terpenes with anticancer properties [33].

Conclusion

This chapter highlights the importance of investigating the structure and activity of sesquiterpenes, as they show positive health effects, especially in anti-inflammatory processes and cancer development.

The most well-known representatives of the sesquiterpenes categories with their activities are listed to emphasize the possibility of studying the connection between their structure and activity to be able to synthesize new derivatives with better activity and bioavailability. Furthermore, given their diversity, a large number of hitherto unknown sesquiterpenes are still being found and isolated from plant sources, which makes this topic still relevant.

Given the large number of different sesquiterpenes found in the plants today, this area has not been sufficiently explored and further research is needed to elucidate the biosynthesis of a larger number of sesquiterpenes and the mechanisms of their action in the field of treatment of inflammation and cancer.

REFERENCES

[1] Modzelewska A, Sur S, Kumar SK, Khan, SR. Sesquiterpenes: Natural Products That Decrease Cancer Growth. *Curr Med Chem Anticancer Agents* 2005; 5:477-499.

[2] Zamani S, Emami SA, Iranshahi M, Rabe, SZT, Mahmoudi, M. Sesquiterpene fractions of Artemisia plants as potent inhibitors of inducible nitric oxide synthase and cyclooxygenase-2 expression. *Iran J Basic Med Sci* 2019; 22:774-780.

[3] Biesalski HK, Dragsted LO, Elmadfa I, Grossklaus R, Müller, M, Schrenk D, et al. Bioactive compounds: definition and assessment of activity. *Nutrition* 2013; 25(11-12):1202-1205.

[4] Ricigliano V, Kumar S, Kinison S, Brooks C, Nybo SE, Chappell J., et al. Regulation of sesquiterpenoid metabolism in recombinant and elicited *Valeriana officinalis* hairy roots. *Phytochemistry* 2016; 125:43-53.

[5] Croteau R, Munck SL, Akoh CC, Fisk HJ, Satterwhite DM. Biosynthesis of the Sesquiterpene Patchoulol from Farnesyl Pyrophosphate in Leaf Extracts of *Pogostemon cablin* (Patchouli): Mechanistic Considerations. *Arch Biochem Biophys* 1987; 256(1):56-68.

[6] Miller DJ, Allemann RK. Sesquiterpene synthases: Passive catalysts or active players? *Nat Prod Rep* 2012; 29:60.

[7] Ashour M, Wink, M, Gershenzon. Biochemistry of terpenoids: monoterpenes, sesquiterpenes and diterpenes. *Annu Plant Rev* 2010; 40:258–303.

[8] Cartayrade A, Schwarz M, Jaun B, Arigoni D. Detection of two independent mechanistic pathways for the early steps of isoprenoid biosynthesis in *Ginkgo biloba*, in *Second Symposium of the European Network on Plant Terpenoid*s. Strasbourg, France; 1994.

[9] Rohmer M, Seemann M, Horbach S, Bringermeyer S, Sahm H. Glyceraldehyde 3-phosphate and pyruvate as precursors of isoprenic units in an alternative non-mevalonate pathway for terpenoid biosynthesis. *J Am Chem Soc* 1996; 118:2564–6.

[10] Vranová E, Coman D, Gruissem, W. Structure and dynamics of the isoprenoid pathway network. *Mol Plant* 2012; 5(2):318–333.

[11] Vranová E, Coman D, Gruissem, W. Network analysis of the MVA and MEP pathways for isoprenoid synthesis. *Annu Rev Plant Biol* 2013; 64:665–700.

[12] Gray JC. Control of isoprenoid biosynthesis in higher plants. *Adv Bot Res* 1987; 14:25–91.

[13] Kleinig H. The role of plastids in isoprenoid biosynthesis. *Annu Rev Plant Physiol Plant Mol Biol* 1989; 40:39–59.

[14] Adam KP, Zapp, J. Biosynthesis of the isoprene units of chamomile sesquiterpenes. *Phytochemistry*, 1998;48:953–959.

[15] Croteau R, Munck SL, Akoh CC, Fisk HJ, Satterwhite DM. Biosynthesis of the Sesquiterpene Patchoulol from Farnesyl Pyrophosphate in Leaf Extracts of *Pogostemon cablin* (Patchouli): Mechanistic Considerations. *Archives of Biochemistry and Biophysics,* 1987;256(1):56-68.

[16] Cane DE. Enzymatic formation of sesquiterpenes. *Chem Rev* 1990; 90:1089–103.

[17] Cane DE. Isoprenoid Biosynthesis: Overview. In: *Comprehensive Natural Products Chem* edited by Barton D, Nakanishi K, Meth-Cohn O, 1998;1-13. Pergamon, Oxford: Elsevier Science.

[18] Chizzola R. Regular Monoterpenes and Sesquiterpenes (Essential Oils). In: *Natural Products* edited by Ramawat KG, Mérillon JM, 2013;2973-3008. Berlin, Heidelberg, Germany: Springer.

[19] Degenhardt J, Köllner TG, Gershenzon, J. Monoterpene and sesquiterpene synthases and the origin of terpene skeletal diversity in plants. *Phytochemistry* 2009; 70:1621–1637.

[20] Gambliel H, Croteau R. Pinene cyclase-I and cyclase-II – two enzymes from Sage (*Salvia officinalis*) which catalyze stereospecific cyclizations of geranyl pyrophosphate to monoterpene olefins of opposite configuration. *J Biol Chem,* 1984; 259:740–748.

[21] Croteau, R. Biosynthesis and catabolism of monoterpenoids. *Chem Rev* 1987; 87:929–954.

[22] Wagschal K, Savage TJ, Croteau R. Isotopically sensitive branching as a tool for evaluating multiple product formation by monoterpene cyclases. *Tetrahedron,* 1991;47:5933–5944.

[23] Vedula LS, Rynkiewicz MJ, Pyun HJ, Coates RM, Cane DE, Christianson DW. Molecular recognition of the substrate diphosphate group governs product diversity in trichodiene synthase mutants. *Biochemistry* 2005; 44:6153–6163.

[24] Awouafack MD, Tane P, Kuete V, Eloff JN. Sesquiterpenes from the Medicinal Plants of Africa. In *Medicinal Plant Research in Africa*, 1st Edition Pharmacology and Chemistry edited by Kuete V, 2013;33-103. Elsevier, Amsterdam, Netherlands.

[25] Durairaj J, Di Girolamo A, Bouwmeester HJ, de Ridder D, Beekwilder J, van Dijk, ADJ. An analysis of characterized plant sesquiterpene synthases. *Phytochemistry* 2019; 158:157–165.

[26] Cordell GA. Biosynthesis of Sesquiterpenes. *Chem Rev* 1976;76(4): 425–460.

[27] Talapatra SK, Talapatra B. Sesquiterpenoids (C15). In *Chemistry of Plant Natural Products*, 2015; 403-468. The Salmon Tower Building, New York, SAD: Springer, Berlin, Heidelberg.

[28] Sotto AD, Mancinelli R, Gullì M, Eufemi M, Mammola CL, Mazzanti G, et al. Chemopreventive Potential of Caryophyllane Sesquiterpenes: An Overview of Preliminary Evidence. *Cancers* 2020; 12:3034.

[29] McAndrew RP, Peralta-Yahya PP, DeGiovanni A, Pereira JH, Hadi MZ, Keasling JD, et al. Structure of a Three-Domain Sesquiterpene Synthase: A Prospective Target for Advanced Biofuels Production. *Structure* 2011; 19:1876–1884.

[30] Son YJ, Kwon M, Ro DK, Kim, SU. Enantioselective microbial synthesis of the indigenous natural product (-)-α-bisabolol by a sesquiterpene synthase from chamomile (*Matricaria recutita*). *Biochem* 2014; 463:239–248.

[31] Sköld M, Karlberg AT, Matura M, Börje A. The fragrance chemical β-caryophyllene—air oxidation and skin sensitization. *Food Chem Toxicol* 2006; 44:538–545.

[32] Croteau R, Munck SL, Akoh CC, Fisk HJ, Satterwhite DM. Biosynthesis of the Sesquiterpene Patchoulol from Farnesyl Pyrophosphate in Leaf Extracts of *Pogostemon cablin* (Patchouli): Mechanistic Considerations. *Arch Biochem Biophys* 1987;256(1): 56-68.

[33] Babaei G, Aliarab A, Abroon S, Rasmi Y, Aziz SGG. Application of sesquiterpene lactone: A new promising way for cancer therapy based on anticancer activity. *Biomed Pharmacother* 2018; 106:239–246.

[34] Simonsen HT. Elucidation of terpenoid biosynthesis in non-model plants utilizing transcriptomic data. *Next Generat Sequenc & Applic*, 2015;2:111.

[35] Perassolo M, Cardillo AB, Busto VD, Giulietti AM, Talou JR. Biosynthesis of Sesquiterpene Lactones in Plants and Metabolic Engineering for Their Biotechnological Production. In: *Sesquiterpene Lactones* edited by Sülsen V, Martino V, 2018;47-91. Springer, Berlin, Heidelberg.

[36] Abu-Izneid T, Rauf A, Ali Shariati M, Khalil AA, Imran M, Rebezov M, et al. Sesquiterpenes and their derivatives-natural anticancer compounds: An update. *Pharmacol Res* 2020; 161: 105165.

[37] Medzhitov R. Origin and physiological roles of inflammation. *Nature* 2008; 454:428–435.

[38] e Sá RCS, Andrade LN, de Sousa DP. Sesquiterpenes from Essential Oils and Anti-Inflammatory Activity. *Nat Prod Commun* 2015;10(10):1767-1774.

[39] Joo JH, Jetten AM. Molecular mechanisms involved in farnesol-induced apoptosis. *Cancer Lett* 2010; 287:123–35.

[40] Pryor WA, Stone K. Oxidants in cigarette smoke: radicals, hydrogen peroxide, peroxynitrate, and peroxynitrite. *Ann N Y Acad Sci* 1993; 686:12–27.

[41] Qamar W, Sultana S. Farnesol ameliorates massive inflammation, oxidative stress and lung injury induced by intratracheal instillation

of cigarette smoke extract in rats: An initial step in lung chemoprevention. *Chem Biol Interact* 2008; 176:79–87.

[42] Jahangir T, Khan TH, Prasad L, Sultana, S. Farnesol prevents Fe-NTA mediated renal oxidative stress and early tumour promotion markers in rats. *Hum Exp Toxicol* 2006; 25:235–242.

[43] Khan R, Sultana S. Farnesol attenuates 1,2-dimethylhydrazine induced oxidative stress, inflammation and apoptotic responses in the colon of Wistar rats. *Chem Biol Interact* 2011;192: 193–200.

[44] Wen J, You KR, Lee SY, Song CH, Kim DG. Oxidative stress-mediated apoptosis: the anticancer effect of the sesquiterpene lactone parthenolide. *J Biol Chem* 2002; 277:38954–38964.

[45] Kamatou GPP, Viljoen AM. A Review of the Application and Pharmacological Properties of a-Bisabolol and a-Bisabolol-Rich Oils. *J Am Oil Chem Soc* 2010; 87:1–7.

[46] Braga PC, Dal Sasso M, Fonti E, Culici M. Antioxidant activity of bisabolol: inhibitory effects on chemiluminescence of human neutrophil bursts and cell-free systems. *Pharmacology* 2009;83: 110-115.

[47] Ghelardini C, Galeotti N, Di Cesare Mannelli L, Mazzanti G, Bartolini A. Local anaesthetic activity of beta-caryophyllene. *Farmaco* 2001;56: 387–389.

[48] Gertsch J, Leonti M, Raduner S, Racz I, Chen JZ, Xie XQ, et al. Beta-caryophyllene is a dietary cannabinoid. *Proc Natl Acad Sci USA* 2008; 105:9099–9104.

[49] Ormeño E, Baldy V, Ballini C, Fernandez C. Production and diversity of volatile terpenes from plants on calcareous and siliceous soils: effect of soil nutrients. *J Chem Ecol* 2008; 34:1219–1229.

[50] Guindon J, Hohmann AG. Cannabinoid CB2 receptors: a therapeutic target for the treatment of inflammatory and neuropathic pain. *Br J Pharmacol* 2008; 153:319-334.

[51] Zheng GQ, Kenney PM, Lam LKT. Sesquiterpenes from clove (*Eugenia caryophyllata*) as potential anticarcinogenic agents. *J Nat Prod* 1992; 55:999–1003.

[52] Passos GF, Fernandes ES, Campos MM, Araujo JG, Pesquero JL, Souza GE, et al. Kinin B1 receptor upregulation after lipopolysaccharide administration: role of proinflammatory cytokines and neutrophil influx. *J Immunol* 2004; 172:1839–1847.

[53] Rocha AC, Fernandes ES, Passos GF, Calixto JB, Campos MM. Assessment of TNF alpha contribution to the functional upregulation of kinin B (1) receptors in the mouse paw after treatment with LPS. *Int Immunopharmacol* 2005; 5:1593–1600.

[54] Campos MM, Leal PC, Yunes RA, Calixto JB. Non-peptide antagonists for kinin B1 receptors: new insights into their therapeutic potential for the management of inflammatory and pain diseases. *Trends Pharmacol Sci* 2006; 27:646–651.

[55] Rogerio AP, Andrade EL, Leite DF, Figueiredo CP, Calixto JB. Preventive and therapeutic anti-inflammatory properties of the sesquiterpene α-humulene in experimental airways allergic inflammation. *Br J Pharmacol* 2009; 158:1074-1087.

[56] Dash GK, Ganapathy S, Suresh P, Panda SK, Sahu SK. Analgesic and anti- inflammatory activity of *Annona squamosa* leaves. *Indian J Nat Prod Resour* 2001;17: 32–35.

[57] Chavan MJ, Wakte PS, Shinde DB. Analgesic and anti-inflammatory activity of caryophyllene oxide from *Annona squamosa* L. bark. *Phytomedicine*, 2010;17: 149–151.

[58] Safahyi H, Sabieraj J, Sailer ER, Ammon, HPT. Chamazulene: an antioxidant-type inhibitor of leukotriene B4 formation. *Planta Med,* 1994;60:410-413.

[59] Xian YF, Li YC, Ip SP, Lin ZX, Lai XP, Su ZR. Anti-inflammatory effect of patchouli alcohol isolated from *Pogostemonis herba* in LPS-stimulated RAW264.7 macrophages. *Exp Ther Med* 2011;2(3): 545–550.

[60] Cordeiro RA, Teixeira CEC, Brilhante RSN, Castelo-Branco DSCM, Paiva MAN, Leite JJG, et al. Minimum inhibitory concentrations of amphotericin B, azoles and caspofungin against Candida species are reduced by farnesol. *Med Mycol* 2013; 51(1):53-59.

[61] Lopes NP, Kato MJ, Andrade EH, Maia JG, Yoshida M, Planchart AR, et al. Antimalarial use of volatile oil from the leaves of *Virola surinamensis* by Waiapi Amazon Indians. *J Ethnopharmacol* 1999;67(3):313-319.

[62] Di Campli E, Di Bartolomeo S, Delli Pizzi P, Di Giulio M, Grande R, Nostro A, et al. Activity of tea tree oil and nerolidol alone or in combination against *Pediculus capitis* (head lice) and its eggs. *Parasitol Res* 2012;111(5):1985-1992.

[63] Medeiros R, Passos GF, Vitor CE, Koepp J, Mazzuco TL, Pianowski LF, et al. Effect of two active compounds obtained from the essential oil of *Cordia verbenacea* on the acute inflammatory responses elicited by LPS in the rat paw. *Br J Pharmacol* 2007; 151:618–627.

[64] Fernandes ES, Passos GF, Medeiros R, Cunha FM, Ferreira J, Campos MM, et al. Anti-inflammatory effects of compounds alpha-humulene and (−)-trans-caryophyllene isolated from the essential oil of *Cordia verbenacea*. *Eur J Pharmacol* 2007; 569:228–236.

[65] Da Silva SL, Figueiredo PM, Yano T. Chemotherapeutic potential of the volatile oils from *Zanthoxylum rhoifolium* Lam leaves. *Eur J Pharmacol* 2007; 8:180-188.

[66] Jin JH, Lee DU, Kim YS, Kim HP. Anti-allergic activity of sesquiterpenes from the rhizomes of *Cyperus rotundus*. *Archives Pharmacal Research*, 2011;34:223-228.

[67] Ornano L, Venditti A, Ballero M, Sanna C, Quassinti L, Bramucci M, et al. Chemopreventive and antioxidant activity of the chamazulene-rich essential oil obtained from *Artemisia arborescens* L. growing on the Isle of La Maddalena, Sardinia, Italy. *Chem Biodivers* 2013;10(8):1464-74.

[68] Kiyohara H, Ichino C, Kawamura Y, Nagai T, Sato N, Yamada H. Patchouli alcohol: in vitro direct anti-influenza virus sesquiterpene in *Pogostemon cablin* Benth. *J Nat Med* 2012; 66:55-61.

[69] Li YC, Xian YF, Ip, SP, Su ZR, Su JY, He JJ, et al. Anti-inflammatory activity of patchouli alcohol isolated from

Pogostemonis herba in animal models. *Fitoterapia*, 2011;82(8):1295-1301.

[70] Muangphrom P, Seki H, Fukushima EO, Muranaka T. Artemisinin-based antimalarial research: application of biotechnology to the production of artemisinin, its mode of action, and the mechanism of resistance of *Plasmodium* parasites. *J Nat Med* 2016; 70:318–334.

[71] Krishna S, Bustamante L, Haynes RK, Staines HM. Artemisinins: their growing importance in medicine. *Trends Pharmacol Sci* 2008;29(10):520–527.

[72] Sun J, Zhang C, Bao YL, Wu Y, Chen ZL, Yu CL, et al. Parthenolide-induced apoptosis, autophagy and suppression of proliferation in HepG2 cells. *Asian Pac J Cancer Prev,* 2014;15:4897–4902.

[73] Khan M, Li T, Ahmad Khan MK, Rasul A, Nawaz F, Sun M, et al. Alantolactone induces apoptosis in HepG2 cells through GSH depletion, inhibition of STAT3 activation, and mitochondrial dysfunction. *Biomed Res Int* 2013; 2013:719858.

[74] Jiang Y, Xu H, Wang J. Alantolactone induces apoptosis of human cervical cancer cells via reactive oxygen species generation, glutathione depletion and inhibition of the Bcl-2/Bax signaling pathway. *Oncol Lett* 2016; 11:4203–4207.

[75] Lohberger B, Rinner B, Stuendl N, Kaltenegger H, Steinecker-Frohnwieser B, Bernhart E, et al. Sesquiterpene lactones downregulate G2/M cell cycle regulator proteins and affect the invasive potential of human soft tissue sarcoma cells. *PLoS One* 2013;8(6): e66300.

[76] Wang Z, Zhao X, Gong X. Costunolide induces lung adenocarcinoma cell line A549 cells apoptosis through ROS (reactive oxygen species)-mediated endoplasmic reticulum stress. *Cell Biol Int* 2015;40(3):289-97.

[77] Zhang C, Lu T, Wang GD, Ma C, Zhou YF. Costunolide, an active sesquiterpene lactone, induced apoptosis via ROS-mediated ER stress and JNK pathway in human U2OS cells. *Biomed Pharmacother* 2016; 80:253–259.

[78] Wu M, Zhang H, Hu J, Weng Z, Li C, Li H, et al. Isoalantolactone inhibits UM-SCC-10A cell growth via cell cycle arrest and apoptosis induction. *PLoS One* 2013;8(9): e76000.

[79] Kabeer FA, Sreedevi GB, Nair MS, Rajalekshmi DS, Gopalakrishnan LP, Kunjuraman S, et al. Antineoplastic effects of deoxyelephantopin, a sesquiterpene lactone from *Elephantopus scaber*, on lung adenocarcinoma (A549) cells. *Chin J Integr Med* 2013; 11:269–277.

[80] Farha AK, Dhanya SR, Mangalam SN, Geetha BS, Latha PG, Remani, P. Deoxyelephantopin impairs growth of cervical carcinoma SiHa cells and induces apoptosis by targeting multiple molecular signaling pathways. *Cell Biol Toxicol* 2014; 30(6):30331–343.

Chapter Reviewed by:

Associate Professor Valentina Pavić, Sub-Separtment of Biochemistry and Plant Ecophysiology, Department of Biology, University of Osijek, Osijek, Croatia.

Assistant professor Jelena Vladić, Department of Biotechnology and Pharmaceutical Engineering, Faculty of Technology Novi Sad, University of Novi Sad, Novi Sad, Serbia.

In: Bioactive Compounds …
Editors: G. Yildiz and G. İzli
ISBN: 978-1-68507-361-9

Chapter 7

From the Plant to the Body: Phytochemicals and Their Bioactivity for Humans

Gabriela Niemeyer Reissig[1,*], André Geremia Parise[1], Douglas Antônio Posso[1], Ádrya Vanessa Lira Costa[1], Lisiane Pintanela Vergara[2], Thiago Francisco de Carvalho Oliveira[1] and Cesar Valmor Rombaldi[2]

[1]Department of Botany, Biology Institute, Federal University of Pelotas - UFPel, Capão do Leão, Rio Grande do Sul, Brazil

[2]Department of Agroindustrial Science and Technology, Eliseu Maciel Agronomy Faculty, Federal University of Pelotas - UFPel, Capão do Leão, Rio Grande do Sul, Brazil

[*] Corresponding Author's Email: gabriela.niemeyer.reissig@gmail.com.

ABSTRACT

Plants are sessile organisms that evolved several mechanisms to cope with the adversities imposed by the environment. One of the strategies to deal with environmental stressors is the production of compounds known as specialized/secondary metabolites. These compounds are known for not being present in all plants and not being directly involved in the growth, development, or reproduction of the organism. In addition to stressful situations, plants also accumulate specialized metabolites in physiological processes, such as fruit ripening. These compounds are also important to fundamental plant processes, such as photosynthesis, respiration, and phenotypic plasticity. However, the human intellect learned to use these substances for other purposes not related at all to plant survival and maintenance. We take advantage of the potential of these compounds and use them to fight or prevent problems associated with several diseases, such as stress. The main goal of this chapter is to link the importance of the phytochemicals in the plant to the bioactive compounds in the human diet. Phenolic compounds, carotenoids, ascorbic acid, alkaloids, glucosinolates, among others, will be discussed. We believe that a broad knowledge of the function of phytochemicals is of the utmost importance for students, professors, and researchers to better understand the main concerns about bioactive compounds. The use of moderate stress techniques to improve the quality of plants in terms of bioactive compounds, as well as studies related to extraction and optimization techniques, food biofortification and enrichment, encapsulation, and nanotechnology for application in food, molecular engineering, plant tissue culture, and studies with unconventional food plants are examples of the importance of cooperation and knowledge from different areas of science. Despite the large number of studies already conducted on phytochemicals and their bioactive capacity, there is still much to be explored.

Keywords: bioactive compounds, health, oxidative stress, plant physiology, secondary metabolites

INTRODUCTION

The importance of plants for life on Earth is unquestionable. They are autotrophic organisms that compose the basis of the food web, being the primary energy suppliers, and supporting the diversity of life. Plants provide food, fuel, shelter, clothing, and the maintenance of oxygen in the atmosphere [1]. The diversity of plants on our planet is astonishing. This huge diversity explains the diversity of phytochemicals produced by them. In addition to compounds that provide nutrition (carbohydrates, lipids, and proteins), plants can synthesize a variety of substances that can vary from species to species, and which have different functions. One of these functions is to protect the plant against the adversities imposed by the environment [2, 3].

An interesting aspect of these compounds is the biological activity they perform. When ingested by animals they can present activities that are the same or quite different from those they exert on the plant. L-ascorbic acid is an example. It presents a pleiotropic behavior, that is, it is a compound with several biological functions. It may act as a vitamin for human nutrition and an antioxidant both for plants and humans [4]. Cisneros-Zevallos [2] argues that oxidative stress seems to be the link between its function in humans and plants, once it is a common redox imbalance among living beings. Plants synthesize chemicals to deal with oxidative stress, protecting cell structures and acclimating. When humans are under oxidative stress the organism reacts through inflammatory processes, and the consumption of some phytochemicals leads to a reduction in the inflammatory processes and associated diseases.

Long before the establishment of scientific thought and methodology, humanity was already aware of the use of different plants for the benefit of human health. Studies in ethnobotany and ethnopharmacology increasingly demonstrate the connection of humankind with plants, showing the close connection of knowledge about plants and culture. Nowadays, much of the traditional and popular knowledge about the use of plants is scientifically supported. Currently, we explore these traditional medicinal plants for a multitude of purposes, one of which is their bioactive properties, acting as antioxidant, anti-inflammatory, antidiabetic, antiemetics, anticancer, antidiarrheal, antibacterial, antifungal, among others [5]. To obtain bioactive effects, plant materials can be consumed directly or through teas and infusions, processing and industrialization of food, cosmetics, fragrances, and supplements.

We already have a vast literature on bioactive compounds. Nonetheless, more studies are still needed to elucidate some issues that still raise many concerns. One of the major issues is the bioavailability and bioaccessibility of these compounds when ingested [6]. Another gap to be filled is the question of the recommended dose, and whether it would be possible to treat chronic diseases with these compounds [2]. In addition, studies aimed at obtaining higher concentrations of these compounds are especially important to make them more accessible to people. Thus, to combine the knowledge of pre-and post-harvest techniques, as well as different research areas, is essential for the advancement of studies on bioactive compounds. In this chapter we write about the main classes of bioactive compounds, briefly reviewing their chemistry, biosynthesis, and bioactive functions in plants and the human body. In addition, we also discuss some means to enhance the obtaining of these compounds and recent studies on the topic.

Bioactive Compounds: From the Plant to the Body

As mentioned before, plants are the basis of our diet, and in addition to nutrition, plants can provide us a plethora of beneficial, medicinal, and recreational substances. Knowledge about the use of plants for their medicinal and bioactive properties goes back to the dawn of human history. There is evidence written on clay tablets that the Sumerians already use plants for medicinal purposes around 5,000 years BC [7]. Long before scientific thought guided our search for knowledge, humanity found ways to be able to take advantage of nature's resources in different ways. With the development of the scientific method and the consequent advance in the various areas of knowledge, it is now possible to understand what is behind the different properties observed in different plants and how we can further explore these resources. For example, today it is known that many of these compounds are produced for the defense of the plant and to deal with oxidative stress, and it is possible to obtain different compounds from different parts of the plant (Figure 1), as well as produce industrialized products with bioactive compounds from plant sources [8]. Also, plant breeding, plant stimulation, and plant nutrition, through different techniques, increase the concentration of these compounds [9, 10]. This strategy is also known as food biofortification [11].

Figure 1. From the plant to the body. Representation of environmental factors responsible for increasing the accumulation of specialized metabolites in plants, followed by ingestion sources and their respective biological effects on human physiology.

Different classes of compounds are responsible for the bioactive properties of plants in our body, such as carotenoids and phenolic compounds. One of the most universal properties of these compounds, both in plants and animals, is as an antioxidant. Since the great oxidation event about 2.4 billion years ago, which culminated in the emergence of aerobic species, living organisms had to deal with the formation of reactive oxygen species (ROS), which can lead to oxidative stress and damage cellular structures [12]. This occurs from a single cell to the largest living beings on our planet, such as whales and redwoods. Now, we can glimpse a little bit of the importance of bioactive compounds for life on Earth. In the following subsections, the main classes of phytochemicals with bioactive properties will be addressed, giving an overview on them and citing some examples of their action on the plant and the human organism.

Phenolic Compounds

Phenolic compounds are secondary metabolites produced by plants and are involved in defense mechanisms against pathogens, parasites, and predators. Due to their chemical structure, they act as an antioxidant in the scavenging of free radicals. In the human body, they act against premature cell aging and susceptibility to infections. In addition, they enhance anti-inflammatory activity [13]. Chemically, phenolic compounds contain an aromatic ring with one or more hydroxyl substituents, along with several other side groups [14]. Phenolic compounds are generally biosynthesized via the shikimic acid pathway [15]. This pathway converts carbohydrate precursors derived from glycolysis and the pentose phosphate pathway into aromatic amino acids [16]. The shikimic acid pathway is present in plants, fungi, and bacteria, but is not found in animals. In plants, the plastids play an important role in the central phenylpropanoids pathway as well for carotenoids [15].

The most abundant class of phenolic compounds in plants is derived from phenylalanine, through the elimination of an ammonia molecule to form cinnamic acid. This reaction is catalyzed by phenylalanine amonialiasis (PAL). PAL activity is increased by environmental factors, such as low levels of nutrients, light (due to its effect on phytochrome), and fungal infection. Subsequent reactions to those catalyzed by PAL lead to the addition of more hydroxyl groups and other substituents.

Trans-cinnamic and p-coumaric acids and their derivatives are simple phenolic compounds called phenylpropanoids, as they contain a benzene ring and a three-carbon side chain. Phenylpropanoids are important basic units for the formation of more complex phenolic compounds [17].

Table 1. Phenolic compounds and their functions in the plant and their bioactivity in humans

Phenolic compounds	Function in the plant	Bioactivity for humans	References
Flavonoids	Cell protection against UV-B radiation excess; modulators of the polar transport of auxin.	Antioxidant; anticancer; anti-inflammatory; antimutagenic.	[17, 21]
Tannins	Plant defense against predation and support in growth regulation.	Antiviral; anticancer, antibacterial; anti-inflammatory; antioxidant.	[22]
Phenolic acids	Plant growth, development and defense.	Antioxidant; anti-inflammatory; antiallergic; antiatherogenic; antimicrobial; anticancer; antidiabetic.	[23, 24]
Tocopherol	Singlet oxygen scavenger in photosystem II.	Antioxidant; anticancer; antihypertensive; nephroprotective; neuroprotective; anti-inflammatory.	[25, 26]

Caffeic and ferulic acid can be released into the soil and inhibit the growth of adjacent plants. Psoralen is a furanocoumarin that has phototoxic effects on herbivorous insects. Lignin, a highly complex phenolic macromolecule, is formed by a highly branched phenyl-propanoid polymer that has primary and secondary functions. Lignin is generally formed by three different phenylpropanoid alcohols (coniferyl, coumaril, and sinapil) which are synthesized from phenylalanine through various derivatives of cinnamic acid. It acts as mechanical support and protection in vegetables. Furthermore, lignification blocks the growth of pathogens and is a frequent response to infection or injury [17].

Phenolic compounds are found mainly in fruits, vegetables, tea, wine, coffee and are responsible for some of the sensory characteristics of plant foods. Bitterness and astringency are characteristic of some phenols [14, 18]. Among plants, flavonoids constitute the largest class of plant phenolics. Structurally, flavonoids are composed of 15 carbons organized in two aromatic rings, connected by a three-carbon chain [18]. Depending on the oxidation state of the heterocyclic chain of pyran, there are different classes of flavonoids: anthocyanins, flavonols, flavones, isoflavones, flavonones, and flavans, with multiple biological effects, such as antioxidant, anti-inflammatory, and antitumor activity, reducing power to capillary fragility and permeability, inhibition of collagen destruction and platelet aggregation. Thus, the intake of flavonoids is associated with longevity and a reduction in the incidence of cardiovascular diseases [19]. Among the flavonoids, anthocyanins stand out. They are glycosides that present in their chemical structure a sugar residue in carbon 3 and play a significant role in preventing or delaying the onset of various diseases due to their antioxidant properties. The anthocyanins found in foods are all derived from the aglycone belonging to three basic pigments: pelargonidin (red), cyanidin (red), and delphinidin (violet) [20]. Table 1 lists some phenolic compounds present in plants, showing their function in the plant and their bioactive properties in the human body.

Carotenoids

Carotenoids are biomolecules synthetised by plants, algae, and fungi, and are responsible for their yellow, orange, and red colors. They are important components of the photosynthetic process, being involved in light energy absorption, redirection, channeling, and releasing. The carotenoids are mainly hydrophobic molecules, and they are divided into two main groups: the hydrocarbon carotenes and their oxygenated derivatives xanthophylls [27, 28]. The biosynthesis pathway of carotenoids starts with isoprene, a five-carbon molecule that is grouped to form a tetraterpenoid, a 40-carbon molecule (eight initial isoprenes). Their colors are related to the number of double bonds in the molecule. A minimum of seven double bonds are necessary to absorb light in the yellow visible specter, and a higher number of these bonds increases the color bands absorbed and turns then red-colored. Similarly, the number of cis and trans isomers influences its color by re-arranging the bonds in the structure.

After the formation of farnesyl pyrophosphate (FPP; 15-carbon) by the bond of three molecules of isoprenes, the FPP is then used to produce geranylgeranyl pyrophosphate (GGPP; 20-carbon). The last step uses two molecules of GGPP to form phytoene (40-carbon). The following steps are responsible to increase the number of double bonds, cyclization, and insertion of oxygenated function. The enzyme phytoene synthase is the most important regulatory enzyme in the biosynthesis pathway. The dehydrogenation of the phytoene proceeds until the obtaining of lycopene, the carotenoid with the greater number of double bonds, presenting the red color. Lycopene is the precursor of the cyclic carotenoids, originating the β-carotene and α-carotene, and then, originate the xanthophylls by oxygenating the structure [28, 29].

Lycopene, β-carotene, and other carotenoids that possess intense colors are important for the survival of different plant species, attracting animals that will consume the fruits and disperse the seeds, and thus securing the spatial distribution of the species. They are also important as a visual sensory attribute for fruits and vegetables, allowing humans to assess, for example, whether some fruit is in the ideal ripening stage for consumption. Regarding xanthophylls, they are important in the photosynthesis process. The xanthophyll cycle acts dissipating the energy excess by transforming zeaxanthin in violaxanthin and vice-versa, attenuating the negative effects of high light conditions, and dissipating the energy that could be used to produce ROS in the photosynthetic process. It is well known that carotenoids act as antioxidant molecules, scavenging the ROS formed by metabolism due to their capacity of accepting and transferring electrons, as well as quenching singlet oxygen. As sessile organisms, plants have evolved to survive by adjusting their metabolism in response to environmental fluctuations, and carotenoids contribute to it [29].

Carotenoids are important both for plants and human health. For human nutrition, they provide the precursor of vitamin A (retinol), being β-carotene one of the most important sources. Zeaxanthin has no vitamin A activity, but together with lutein play a crucial role in the prevention of degenerative diseases, beyond mitigating and promoting free radical scavenging. They are also important in the prevention of chronic diseases. Also, it presents anticancer properties, mainly due to its anti-inflammatory capacity [27, 29]. Humans cannot produce these compounds, so fruits and vegetables are the main sources. Table 2 lists some carotenoids present in plants, showing their function in the plant and their bioactive properties in the human body.

Table 2. Carotenoids and their functions in the plant and their bioactivity in humans

Carotenoids	Function in the plant	Bioactivity for humans	References
β-carotene	Coloring flowers and fruit to attract pollinators and dispersers; ROS scavenging	ROS scavenging; Preventing Parkinson's disease; anti-inflammatory properties.	[29, 30, 31]
Zeaxanthin	Xanthophyll cycle – High light protection; ROS scavenging.	Preventing diabetic retinopathy; ROS scavenging.	[29, 32]
Lutein	Responsible to act in the protection of the photosynthetic machinery; ROS scavenging.	Preventing stroke, atherosclerosis, and diabetic retinopathy; eye protection; ROS scavenging; anti-inflammatory activity.	[29, 33, 34, 35]
Lycopene	Coloring flowers and fruit to attract pollinators and dispersers; ROS scavenging; high-light protection	Anti-cancer; ROS scavenging; photo-oxidative damage protection.	[29, 36, 37]

Glucosinolates

Glucosinolates (GSLs) are a group of plant metabolites that contain nitrogen and sulfur in their structures. In short, they are a thiohydroximate with a sulfate ester group at the oxygen and a glucose residue at the sulfur. The rest of the GSLs molecules are aglycone. It is the variations in the aglycone (side chain or "R-group") that confer the diversity of biological functions of GSLs [38, 39]. Regarding biosynthesis, GLSs are formed from amino acids and can be divided into three groups (aliphatic, indole, and aromatic), depending on the amino acid from which they originate. Some precursors are phenylalanine, tryptophan, and methionine. In a nutshell, for GSLs biosynthesis it is necessary side-chain elongation of amino acids, formation of the core glucosinolate structure, and secondary modifications of the sidechain [38, 39]. GSLs are responsible for the characteristic flavor and aroma of brassicaceous vegetables [40].

The Brassicaceae family has great importance for food and research. Mustard (*Brassica juncea* L.), radish (*Raphanus sativus* L.), cauliflower (*Brassica oleracea* var. botrytis L.), broccoli (*Brassica oleracea* L.), among others, are examples of brassicaceous vegetables. It is worthy to mention a special representative of the family, the *Arabidopsis thaliana*, one of the most famous plants among plant researchers because it is a model organism for many studies due to its small size, rapid life cycle, and its full-sequenced genome [41]. One of the characteristics of this family is the presence of GSLs, which are directly involved in plant defense.

The GSLs and their hydrolysis products may act as a defense against pathogens and insects. The glucosinolate–myrosinase system, also known as the "mustard oil bomb", is a defense mechanism against herbivory. When a plant is under herbivory attack, tissue disruption occurs and the GSLs in the vacuole encounter the enzyme myrosinase, resulting in the hydrolysis and production of derived compounds such as isothiocyanate, which in turn act as a toxic compound for several herbivores [38, 40]. In addition to biotic stress, plants may increase GSLs content when subjected to abiotic stress factors. In a study with drought stress in Chinese cabbage (*Brassica rapa*), the authors observed an increase in total GSLs content that was higher in the second week of water shortage when compared to the third week and the control treatment [42]. GSLs can act as stress signaling molecules, triggering defense responses against drought stress. Regarding high salinity conditions, accumulation of salt ions and the water imbalance in the plant are consequences of this adversity. The increase in GSLs levels may be involved in the plant osmotic regulation [38]. Also, GSLs can act as antioxidants in plants, scavenging reactive oxygen species [43].

Table 3. Glucosinolates and their functions in the plant and their bioactivity in humans

Glucosinolates and its hydrolysis products	Function in the plant	Bioactivity for humans	References
Isothiocyanates	Defense against insects and antimicrobial activity against pathogenic bacteria and fungi.	Anti-inflammatory; anti-cancer; antibacterial and antifungal activity.	[38, 44, 47]
Gluconapin (3 butenyl -glucosinolate)	Defense against bacterial pathogens.	Anti-inflammatory and antidiabetic activity.	[38, 48]
Glucobrassicanapin (4-pentenyl-glucosinolate)	The increment in content is associated with plant response to drought stress.	Antioxidant activity.	[42, 44]
Sinigrin (2-propenyl-glucosinolate)	Controlling soil-borne plant pests (biofumigation).	Anti-cancer, antibacterial; antifungal; antioxidant; anti-inflammatory; wound healing properties.	[49]

When consumed, GSLs present antioxidant, antidiabetic, antitumor, and antimicrobial bioactive properties, among others [44]. Volatile degradation products of GSLs, such as isothiocyanates, are the major compounds related to microbial activity. Most of the studies related to microbial inhibition in foods are conducted with allyl isothiocyanate, due to its natural origin and low toxicity [45]. In addition to the antimicrobial activity, isothiocyanates also present neuroprotective effects in neuronal cells. A remarkable characteristic of several phytochemicals, such as isothiocyanates, is the ability to present more than just one biological activity. A single molecule from a specific GSL can act as an antioxidant and anti-inflammatory in the plant and human physiology [46]. Table 3 lists some GLSs present in plants, showing their function in the plant and their bioactive properties in the human body. Next, we will discuss other plants' N-containing compounds, the alkaloids.

Alkaloids

Alkaloids are roughly defined as pharmacologically alkaline compounds which contain at least one nitrogen atom. These substances are produced by a variety of different organisms like plants, fungi, bacteria, and even animals [50]. Alkaloid is a word borrowed from the Arabic ٱلْقِلْي (*al-qalī*), which means "ashes of plants", notably of saltwort. It refers to the basic soda ashes produced from the combustion of these plants [51]. Due to the bioactive properties of alkaloids, they were used since prehistory by humans for a variety of uses such as medicinal, ritualistic, and divinatory [52, 53].

Among all the natural sources of alkaloids, plants are the main producers [54]. Because of their intricate structures, their synthesis is extremely complex and consequently hard to elucidate. However, many experiments have suggested that the precursors of alkaloids are L-amino acids such as L-tryptophan, the precursor of ajmalicine [55]. These precursors can be employed in combination or not with other compounds like terpenoids, steroids, and secoiridoids [54].

It seems that the primary role of alkaloids in plants is as defensive mechanisms. This thesis is largely supported by the many poisonous effects that alkaloids have on animals and other microorganisms. For example, the most consumed psychotropic drug in the world, caffeine, is strongly related to biotic and abiotic stresses in plants. Its presence in seeds and leaves reduces herbivory by inhibiting phosphodiesterases, which affects cAMP metabolism [56]. In the human brain, on the other side, caffeine acts as an antagonist of adenosine, binding to adenosine receptors in the neurons and acting as a stimulating of the central nervous system (CNS). In low doses, these effects are normally desired, but excessive caffeine can lead to various medical disorders such as insomnia, tachycardia and heart arrhythmia, anxiety disorders, and even chemical dependence [57].

Nicotine, another highly consumed alkaloid despite its notorious harmful effects on human health, is a well-known defense metabolite in plants. After being consumed, it inhibits acetylcholine receptors in the CNS of herbivores with lethal consequences [58]. This is why nicotine is one of the oldest-known insecticides [58]. In an experiment with *Manduca sexta* caterpillars and *Diabrotica undecimpunctata* beetles, the caterpillars developed better in transformed *Nicotiana attenuata* plants with 95% less nicotine, and the beetles preferred the transformed plants over the wild type. Besides, in a field experiment, the transformed plants underwent 3-fold more herbivory damage than the control plants [59].

Morphine, notoriously produced by the opium poppy (*Papaver somniferum*), is one of the oldest-known medicines in the world, as it is a powerful analgesic and narcotic. In the plant, however, it has a defensive role. Upon physical damage of the poppy's capsules and leaves, morphine is rapidly oxidized by the enzyme bismorphine-forming peroxidase in the presence of H_2O_2, which generates the dimer bismorphine. This molecule is responsible for stiffening pectins in the cell wall and protecting the plant from wall-degrading enzymes such as pectinase, which are released by pathogens during infection. Therefore, the production of bismorphine is suggested as the earliest defense reaction against pathogens in poppies [60].

Another example of an important alkaloid is the N, N-dimethyltryptamine (DMT), which occurs naturally in a variety of tropical species such as *Mimosa tenuiflora*, *Diplopterys cabrerana*, and *Psychotria viridis*. It is not clear which is the role of DMT within these plants, but since it strongly affects the CNS, perhaps it has a defensive role like most of the naturally occurring alkaloids. For example, high concentrations of DMT and 5-MeO DMT in *Phalaris aquatica* have been associated with its toxicity to livestock [61]. In humans, however, these plants have been used for centuries in shamanic rituals of Amazonian indigenous peoples because of their powerful entheogenic effects [62]. Besides, the potentiality for the clinical use of DMT has called attention because of the promising results of recent research. DMT seems to be effective for treating neuropsychological disorders like depression and anxiety, sometimes after a single dose [62].

The tropane alkaloids are also important. For example, scopolamine and others are present in *Datura*, *Hyoscyamus*, *Atropa,* and *Scopolia*, and act in plant defense. In animals, they act as parasympatholytic that competitively antagonizes acetylcholine. Anticholinergics are generally used as mydriatics and for controlling gastric acidity, as well as to slow gut motility and prevent vomiting. Besides their medicinal use, they possess hallucinogenic and poisonous properties. In applied medicine, they are used to relieve the symptoms of Parkinson's disease [63].

Table 4. Alkaloids and their functions in the plant and their bioactivity in humans

Alkaloid	Function in the plant	Bioactivity for humans	References
Caffeine	Herbivory defense	CNS stimulating	[56, 57]
DMT	Herbivory defense	Ritualistic, therapeutical	[61, 62]
Morphine	Intermediary of bismorphine (defense)	Analgesic, narcotic	[60]
Nicotine	Herbivory defense	Recreational	[58, 59]
Scopolamine	Herbivory defense	Anticholinergic	[63]

In summary, alkaloids are molecules produced by the secondary metabolism of plants mainly for defense against herbivory. There is an enormous variety of alkaloids that affect in many ways the metabolism of herbivores to deter grazing. However, for millennia humans have hijacked these molecules and exploited their effects for medication, ritualistic and recreational purposes. About 9% of all plant genera worldwide have alkaloid-accumulating species, several of them unknown by science [54]. Many medications of the future are still hidden in the jungles and forests of the world.

Ascorbic Acid

Ascorbic acid (Asc) is essential for mammals and plants. In addition to being a critical cofactor for several enzymes, it is also recognized as a bioactive compound for its role in different biological functions. However, humans and other primates are unable to biosynthesize Asc due to a mutation that resulted in the loss of function of the enzyme L-gulono-1,4-lactone oxidase, which catalyzes the final step in Asc biosynthesis [64]. Asc can be found in vivo in its reduced (ascorbic acid) or oxidized (dehydroascorbic acid, DHA) form [64]. In humans, Asc participates in the synthesis of different compounds (carnitine and collagen), conversion of cholesterol to bile acid, increase in bioavailability and absorption of metals, reduction of LDL oxidation, reduction of blood pressure and risk of cardiac ischemia, prevention of cancer, among others [65]. Scurvy, a disease caused by severe vitamin C deficiency in humans, can be treated only through a restored intake of vitamin C [66].

Asc biosynthesis in plants occurs in the mitochondria through several putative routes. The Smirnoff-Wheeler pathway, also known as the D-mannose/L-galactose pathway, is the most elucidated [67]. In this pathway, D-glucose is converted to D-glucose 6-phosphate by the enzyme hexokinase. The D-glucose 6-phosphate is converted to GDP-D-mannose in four stages. Then, GDP-D-mannose is converted to GDP-L-galactose which is subsequently converted to L-galactose and, finally, to L-galactone-1,4-lactone, which in turn converted to ascorbic acid through L -galactone-1,4-lactone dehydrogenase. A second route involves cell wall pectins, where methyl-galacturonate is initially converted to L-galactonate and then undergoes the action of the enzyme aldono lactonase, forming L-galactone-1,4-lactone, which in turn is finally used in ascorbate synthesis (Smirnoff et al., 2001). Other pathways indicate the formation of L-gulono-1,4-lactone (from GDP-D-mannose) via L-gulose [69] and from myo-inositol [70].

The antioxidant role of Asc prevents the accumulation of ROS and, consequently oxidative damage, increasing plant tolerance to stress [66, 71]. The ascorbate-glutathione cycle in organelles that possess active redox activities, such as mitochondria and peroxisomes, is well reported. Ascorbate peroxidases (APXs) are heme-containing enzymes known to dismute H_2O_2 in water and molecular oxygen, using Asc as an electron source [72]. APX1, expressed in Arabidopsis roots, catalyzes the detoxification of H_2O_2 using Asc as an electron source [73]. Most of the cellular Asc (~90%) is in the cytoplasm. However, although in the apoplast its concentration is minimal (in the milli-molar range), it has an important function in the primary line of defense against external oxidants [74, 75].

In addition to ROS elimination, Asc also protects the chloroplasts against photoinhibition of photosynthesis [76]. The interaction of Asc with zeaxanthin protects light-capturing complexes from high luminosity and regulates the photoinhibition of photosynthesis by thermal dissipation [77]. Also, ascorbate can act as an optional electron donor for the photosynthetic electron transport chain, reversing photoinactivation [78] and regulating phytohormone synthesis during abiotic stresses [79]. As observed, ascorbate is a multifunctional and versatile metabolite playing an essential regulatory role in the main physicochemical processes in plants.

Other Compounds

In addition to the above-mentioned compounds, other molecules present bioactive properties, such as betalains and terpenoids. Some proteins and peptides also present bioactivity, and there are several studies in the literature on their bioactive properties. Some bioactive proteins and peptides are produced by plants as defense mechanisms.

Cereal and legumes are widely studied due to their importance for human nutrition, and they are sources of bioactive proteins and peptides [80]. The mung bean [*Vigna radiata* (L.)] is well known for its detoxification bioactivities. Its seeds are particularly rich in protein. The nsLTP and mungin protein found in the seeds possess anti-pathogenic activity against *Staphylococcus aureus* and *Fusarium oxysporum*, among others [81]. So far, the main groups of bioactive compounds and their functions in plants and humans have been described, as well as some proteins that also have these properties. To further explore these compounds, researchers can use different techniques to improve the production of plant bioactive compounds. In the next section, some of these techniques will be discussed.

How to Improve the Production of Plant Bioactive Compounds?

Knowing that phytochemicals are important for plant defense and have different bioactive functions in the organism, it is to be expected that scientists and researchers want to take full advantage of the benefits promoted by these compounds. One way to do this is to stimulate the plant to produce more through moderate stress or applying elicitor agents. Moderate water deficit can increase phytochemicals content without affecting crop productivity. The phytohormone abscisic acid (ABA) has its accumulation increased in water deficit conditions. This molecule is related to the stomatal opening and closing mechanism, being important in the plant's defense mechanism against abiotic stresses. Crosstalk between ABA, Ca^{2+}, CDPK, and phenylpropanoids is a strong hypothesis. Stress responses involve the cellular uptake of ABA and Ca^{2+} and an increase in ABA biosynthesis. Increases in free ABA induce the uptake of Ca^{2+} and the altered expression of stress related genes, including transcription factors (MYB), related to the synthesis of phenylpropanoids. The higher content of Ca^{2+} is perceived by CDPKs resulting in the phosphorylation cascade involving proteins from the ABA signaling, sugar synthesis, transcriptional factors, and synthesis of phenylpropanoids [3]. In the human body, studies suggest its bioactivity against type II diabetes, obesity, and hypertension associated with atherosclerosis [9]. Elicitors are any physical or chemical agent capable of inducing or stimulating defense response in plant cells. Together with plant cell culture, elicitors represent an excellent alternative to improve bioactive production. It allows controlled conditions, feasibility in the manipulation of culture conditions for higher productivity, and the use of simpler and more economical processes to recover the bioactive compounds [10]. Another possibility is through genetic transformation. Overexpression of genes, for example, can lead to the accumulation of several bioactive compounds. Studies with *Withania somnifera* have shown that the overexpression of genes, such as squalene synthase-related genes, enhance the production of bioactive compounds, in this case, withanolide [82]. Food biofortification in bioactive compounds and nutrients is a priority [11]. Here, a brief discussion was given on what can be done to increase the availability of these compounds in plants. In the following topic, we will briefly discuss some recent studies on plant bioactive compounds.

What Is New about Studies on Bioactive Compounds?

Techniques for extracting bioactive compounds in plants are targets of constant research, always looking for more efficient methods, with less environmental impact and low cost. Some examples of novel extraction technologies applied in the recovery of bioactive compounds are pulsed electric field-assisted extraction (PEF), microwave-assisted extraction (MAE), enzyme-assisted extraction (EAE), and ultrasound-assisted extraction (UAE). There is also subcritical water and supercritical carbon dioxide extraction, which are environmentally friendly techniques.

Examples of compounds already extracted with these techniques are carotenoids, phenolic compounds, withanolides, and cannabinoids [83]. In addition to the extraction techniques, the search for increasingly efficient, non-toxic, natural, and renewable solvents is constant. For example, natural deep eutectic solvents (NADES) appear as good candidates for extraction solvents. They are products of the interactions of organic acids and sugars through hydrogen bonds and Van der Waals interactions. In a study with olive leaves, considered an agricultural waste rich in bioactive compounds, NADES combined with ultrasound-assisted extraction were used. The author tested different combinations of NADES, being the glucose–fructose–water the most efficient to extract oleuropein and caffeic acid [84].

Another major problem in food science and technology is the susceptibility of bioactive compounds to environmental factors and processing, as well as to the conditions of gastrointestinal digestion. One solution to this problem is encapsulation technology. There are several techniques for micro and nano encapsulating particles, some examples are: spray drying, freeze-drying, emulsification, liposomes, electro-spinning, and electrospraying. In food products, phenolic compounds can be added in the encapsulated form in beverages, confectionaries, and dairies [85]. Besides the new techniques to extract and deliver bioactive compounds, studies aimed at the discovery of new plants and new compounds with bioactive potential should also be encouraged, as well as studies to understand its medicinal properties and underlying mechanisms in human physiology, so that we can increasingly treat diseases with natural compounds.

CONCLUSION

Plants and humans do not live with nutritious molecules alone! As we have seen, we need other compounds that provide health benefits to have a better quality of life. Despite the enormous structural difference between plants and humans, we share several physiological aspects that are common to most living organisms. Countless molecules that are produced by plants for their benefit also end up triggering a response in our bodies. Studies focused on plant bioactive compounds should address different topics, such as identification of compounds, pharmacological use, accessibility, availability, obtaining, extracting, among others. In addition, researchers must be aware of the importance of environmental preservation and loss of biodiversity, which directly affects these studies. In the end, everything is connected. Therefore, interdisciplinary thinking in research groups is so important. With a wider approach, we can glimpse the most appropriate answers and solutions to the problems of our society.

ACKNOWLEDGMENTS

This study was supported in part by the Coordenação de Aperfeiçoamento de Pessoal de Nível Superior, Brazil (CAPES), Finance Code 001. Authors are also grateful to Conselho Nacional de Desenvolvimento Científico e Tecnológico (CNPq) for the financial support provided (401722/2016-3; 301187/2018-5).

The authors are grateful to the expert reviewers Professor Dr. Junior Borella and Dr[a]. Simone Ribeiro Lucho for their advice on the chapter. J. B. (Lic, MSc, PhD), Institute of Biological Sciences, Federal University of Rio Grande - FURG, Rio Grande, Rio Grande do Sul, Brazil; S. R. L. (Lic, MSc, PhD), Department of Botany, Biology Institute, Federal University of Pelotas – UFPel, Capão do Leão, Rio Grande do Sul, Brazil.

REFERENCES

[1] Knapp S. People and plants: The unbreakable bond. *Plants People Planet* 2019;1:0–26.

[2] Cisneros-Zevallos L. The power of plants: how fruit and vegetables work as source of nutraceuticals and supplements. *Int J Food Sci Nutr* 2020.

[3] Vighi IL, Crizel RL, Perin EC, Rombaldi CV, Galli V. Crosstalk During Fruit Ripening and Stress Response Among Abscisic Acid, Calcium-Dependent Protein Kinase and Phenylpropanoid, *Crit Rev Plant Sci* 2019;38:99-116.

[4] Akram NA, Shafiq F, Ashraf M. Ascorbic acid-a potential oxidant scavenger and its role in plant development and abiotic stress tolerance. *Front Plant Sci* 2017;8:613.

[5] Ghorbanpour M, Hadian, J, Nikabadi, S, Varma, A. Importance of Medicinal and Aromatic Plants in Human Life. *Medicinal Plants and Environmental Challenges.* Switzerland: Springer International Publishing, 2017.

[6] Barba FJ, Mariutti LRB, Bragagnolo N, Mercadante AZ, Barbosa-Cánovas GV, Orlien V. (2017). Bioaccessibility of bioactive compounds from fruits and vegetables after thermal and nonthermal processing. *Trends Food Sci Technol* 2017;67:195-206.

[7] Inoue M, Hayashi S, Craker LE. Role of Medicinal and Aromatic Plants: Past, Present, and Future. *Pharmacognosy - Medicinal Plants.* London: IntechOpen, 2019.

[8] Reissig GN, Rombaldi CV. Brazilian native fruits: From a plurality of examples and bioactive compounds to the most diverse potentialities. *Bioactive Compounds: Sources, Properties and Applications.* Hauppauge: Nova Science, 2017.

[9] Ripoll J, Urban L, Staudt M, Lopez-Lauri F, Bidel LPR, Bertin N. Water shortage and quality of fleshy fruits—making the most of the unavoidable. *J Exp Bot* 2014;65: 4097–4117.

[10] Narayani M, Srivastava S. Elicitation: a stimulation of stress in in vitro plant cell/tissue cultures for enhancement of secondary metabolite production. *Phytochem Rev* 2017;16:1227–1252.

[11] Bouis HE, Saltzman A. Improving nutrition through biofortification: A review of evidence from HarvestPlus, 2003 through 2016. *Glob Food Sec* 2017;12:49-58.

[12] Turkan I. Emerging roles for ROS and RNS – versatile molecules in plants. *J Exp Bot* 2017;68: 4413–4416.

[13] Liu RH. Health benefits of fruit and vegetables are from additive and synergistic combinations of phytochemicals. *Am J Clin Nutr* 2003;78:517–520.

[14] Chandrasekara A. Phenolic Acids. *Enc Food Chem* 2019;3:535-545.

[15] Sun X, Li X, Shen X, Wang J, Yuan Q. Review, Recent advances in microbial production of phenolic compounds. *Chin J Chem Eng* 2021;30:54-61.

[16] Hermann K, Weaver LM. Lignin formation in plants. The dilemma of linkage specificity. *Plant Physiol* 2001;126:1351-1357.

[17] Taiz L, Zeiger E. *Plant Physiology.* 3Ed. Porto Alegre: Editora Artmed, 2004.

[18] Alara OR, Abdurahman NH, Ukaegbu CI. Extraction of phenolic compounds: A review. *Curr Res Food Sci* 2021;4:200-214.

[19] Filho DW, Silva EL, Boveris A. Antioxidant flavonoids of plants and food: importance and therapeutical perspectives. In: Yunes, R. A. e Calixto, J. B. *Medicinal Plants: under the optics of modern medicinal chemistry,* Chapecó: Agros, 2011.

[20] Castaneda-ovando A, Pacheco-hernandez MDEL, Paezhernandez ME, Rodriguez JÁ, Galan-vidal CA. Chemical studies of anthocyanins: A review. *Food Chem,* 2009;11:859-871.

[21] Panche AN, Diwan AD, Chandra SR. Flavonoids: an overview. *J Nutr Sci* 2016;5:e47.

[22] Das AK, Islam MN, Faruk MO, Ashaduzzaman M, Dungani R. Review on tannins: Extraction processes, applications and possibilities. *S Afr J Bot* 2020;135:58-70.

[23] Kumar N, Goel N. Phenolic acids: Natural versatile molecules with promising therapeutic applications. *Biotechnol Rep* 2019;24: e00370.

[24] Anantharaju PG, Gowda PC, Vimalambike MG, Madhunapantula SV. An overview on the role of dietary phenolics for the treatment of cancers. *Nutr* 2016;15:99.

[25] Mocchegiani E, Costarelli L, Giacconi R, Malavolta M, Basso A, Piacenza F, Ostan R, Cevenini E, Gonos ES, Franceschi C, Monti D. Vitamin E–gene interactions in aging and inflammatory age-related diseases: Implications for treatment. A systematic review. *Ageing Res Rev* 2014;14:81–101.

[26] Rashid Khan M, Ahsan H, Siddiqui S, Siddiqui WA. Tocotrienols have a nephroprotective action against lipid-induced chronic renal dysfunction in rats. *Ren Fail* 2014;37:136 – 143.

[27] Rodriguez-Amaya DB. Update on natural food pigments - A mini-review on carotenoids, anthocyanins, and betalains. *Food Res Int* 2019;124:200-205.

[28] Zhao M-H, Li X, Zhang X-X, Zhang H, Zhao X-Y. Mutation Mechanism of Leaf Color in Plants: A Review. *Forests* 2020;11: 851.

[29] Jain A, Sirisha VL. Algal Carotenoids: Understanding Their Structure, Distribution and Potential Applications in Human Health. In: *Encyclopedia of Marine Biotechnology*. New York: Wiley, 2020.

[30] Etminan M, Gill SS, Samii A. Intake of vitamin E, vitamin C, and carotenoids and the risk of Parkinson's disease: a meta- analysis. *Lancet Neurol* 2005;4:362–365.

[31] Kaliappan G, Nagarajan P, Moorthy R, Kalai Gana Selvi S, Avinash Raj T, Mahesh Kumar J. Ang II induce kidney damage by recruiting inflammatory cells and up regulates PPAR gamma and Renin 1 gene: effect of beta carotene on chronic renal damage. *J Thromb Thrombolysis* 2013;36:277–285.

[32] Kowluru RA, Menon B, Gierhart DL. Beneficial effect of zeaxanthin on retinal metabolic abnormalities in diabetic rats. *Invest Ophthalmol Vis Sci* 2008;49:1645–1651.

[33] Xu XR, Zou ZY, Xiao X, Huang YM, Wang X, Lin XM. Effects of lutein supplement on serum inflammatory cytokines, apoE and lipid proles in early atherosclerosis population. *J Arterioscler Thromb* 2013;20:170–177.

[34] Li SY, Fung FK, Fu ZJ, Wong D, Chan HH, Lo AC. Anti-inflammatory effects of lutein in retinal ischemic/hypoxic injury: in vivo and in vitro studies. *Invest Ophth Vis Sci* 2012;53: 5976–5984.

[35] Agarwal M, Parameswari RP, Vasanthi HR, Das DK. Dynamic action of carotenoids in cardio protection and maintenance of cardiac Health. *Molecules* 2012;17:4755–4769.

[36] Ribaya-Mercado JD, Garmyn M, Gilchrest BA, Russell RM. Skin lycopene is destroyed preferentially over h-carotene during ultraviolet irradiation in humans. *J Nutr* 1995;125:1854–1859.

[37] Heinrich U, Gärtner C, Wiebusch M, Eichler O, Sies H, Tronnier H, Stahl W. Supplementation with beta-carotene or a similar amount of mixed carotenoids protects humans from UV-induced erythema. *J Nutr* 2003;133:98–101.

[38] Sánchez-Pujante PJ, Borja-Martínez M, Pedreño MA, Almagro L. Biosynthesis and bioactivity of glucosinolates and their production in plant in vitro cultures. *Planta* 2017;246:19–32.

[39] Blažević I, Montaut S, Burčul F, Olsen CE, Burow M, Rollin P, Agerbirk N. Glucosinolate structural diversity, identification, chemical synthesis and metabolism in plants. *Phytochemistry,* 2020;169:112100.

[40] Chhajed S, Mostafa I, He Y, Abou-Hashem M, El-Domiaty M, Chen S. Glucosinolate Biosynthesis and the Glucosinolate–Myrosinase System in Plant Defense. *Agronomy* 2020;10:1786.

[41] Marchioni I, Martinelli M, Ascrizzi R, Gabbrielli C, Flamini G, Pistelli L, Pistelli L. Small Functional Foods: Comparative Phytochemical and Nutritional Analyses of Five Microgreens of the Brassicaceae Family. *Foods* 2021;10:427.

[42] Shawon RA, Kang BS, Lee SG, Kim SK, Lee HJ, Katrich E, Gorinstein S, Ku YG. Influence of drought stress on bioactive compounds, antioxidant enzymes and glucosinolate contents of Chinese cabbage (Brassica rapa). *Food Chem* 2020;308:125657.

[43] Almuhayawi MS, AbdElgawad H, Jaouni SK, Selim S, Hassan AHA, Khamis G. Elevated CO2 improves glucosinolate

metabolism and stimulates anticancer and anti-inflammatory properties of broccoli sprouts. *Food Chem* 2020;328:127102.

[44] Sikorska-Zimny K, Beneduce L. The glucosinolates and their bioactive derivatives in Brassica: a review on classification, biosynthesis and content in plant tissues, fate during and after processing, effect on the human organism and interaction with the gut microbiota. *Crit Rev Food Sci Nutr,* 2020;25:1-28.

[45] Saladino F, Bordin K, Luciano FB, Franzón MF, Mañes J, Meca G. Antimicrobial Activity of the Glucosinolates. J.-M. Mérillon, K.G. Ramawat (eds.), *Glucosinolates. Reference Series in Phytochemistry.* Switzerland: Springer International Publishing, 2017.

[46] Angeloni C, Hrelia S, Malaguti M. Neuroprotective Effects of Glucosinolates. J.-M. Mérillon, K.G. Ramawat (eds.), Glucosinolates. *Reference Series in Phytochemistry.* Switzerland: Springer International Publishing, 2017.

[47] Textor S, Gershenzon J. Herbivore induction of the glucosinolate–myrosinase defense system: major trends, biochemical bases and ecological significance. *Phytochem Rev* 2009;8:149–170.

[48] Guo Q, Sun Y, Tang Q, Zhang H, Cheng Z. Isolation, identification, biological estimation, and profiling of glucosinolates in Isatis indigotica roots. *J Liq Chromatogr Relat Technol* 2020;43: 645-656.

[49] Mazumder A, Dwivedi A, Du Plessis J. Sinigrin and Its Therapeutic Benefits. *Molecules* 2016;21: 416.

[50] Roberts MF, Wink M. *Alkaloids: biochemistry, ecology, and medicinal applications.* New York: Springer Science+Business Media, 1998.

[51] *OALD: Oxford Oxford Advanced Learner's Dictionary.* Alkali. (2021). Accessed April 16, 2021. https://www.oxfordlearners dictionaries.com/definition/english/alkali?q=alkali.

[52] McKenna DJ, Towers GHN, Abbott F. Monoamine oxidase inhibitors in South American hallucinogenic plants: tryptamine and

β-carboline constituents of ayahuasca. *J Ethnopharmacol,* 1984;10: 195–223.

[53] Uchtenhagen AA. Heroin maintenance treatment: From idea to research to practiced. *Drug Alcohol Rev,* 2011;30:130–137.

[54] Kutchan TM, Gershenzon J, Møller BL, Gang DR. Natural Products. In: Buchanan BB, Gruissen W, Jones RL (eds). *Biochemistry & Molecular Biology of Plants.* Chichester: John Wiley & Sons, 2015.

[55] Liu W, Chen R, Chen M, Zhang H, Peng M, Yang C, Ming X, Lan X, Liao Z. Tryptophan decarboxylase plays an important role in ajmalicine biosynthesis in Rauvolfia verticillata. *Planta* 2012;236:239–250.

[56] Kim Y-S, Choi Y-E, Sano H. Stimulation of defense system by caffeine production in planta. *Plant Signal Behav* 2010;5:489–493.

[57] Winston AP, Hardwick E, Jaberi N. Neuropsychiatric effects of caffeine. *Adv Psychiatr Treat* 2005;11: 432–439.

[58] Millar NS, Denholm I. Nicotinic acetylcholine receptors: targets for commercially important insecticides. *Invertebr Neurosci* 2007;7:53–66.

[59] Steppuhn A, Krock B, Halitschke R, Baldwin IT. Nicotine's defensive function in nature. *PLOS Biol,* 2004;2:e382.

[60] Morimoto S, Suemori K, Moriwaki J, Taura F, Tanaka H, Aso M, Tanaka M, Suemune H, Shimohigashi Y, Shoyama Y. Morphine metabolism in the opium poppy and its possible physiological function: biochemical characterization of the morphine metabolite, bismorphine. *J Biol Chem* 2001;276:38179–38184.

[61] Read E, Reddy P, Rendell D, Rochfort. Changes in field concentrations of five phalaris alkaloids and their association with toxicity in pastures of Victoria, Australia. *Crop Pasture Sci* 2020;71:389–400.

[62] Palhano-Fontes F, Barreto D, Onias H, et al. Rapid antidepressant effects of the psychedelic ayahuasca in treatment-resistant depression: A randomized placebo-controlled trial. *Psychol Med* 2019;49:655-663.

[63] Palazón J, Navarro-Ocaña A, Hernandez-Vazquez L, Mirjalili MH. Application of Metabolic Engineering to the Production of Scopolamine. *Molecules* 2008;13:1722-1742.

[64] Zhitkovich A. Nuclear and Cytoplasmic Functions of Vitamin C. *Chem Res Toxicol* 2020;33:2515-2526.

[65] Gallie D. L-Ascorbic acid: a multifunctional molecule supporting plant growth and development. *Scientifica* 2013:1-24.

[66] Xiao M, Li Z, Zhu L, Wang J, Zhang B, Zheng F, Zhao B, Zhang H, Wang Y, Zhang Z. The Multiple Roles of Ascorbate in the Abiotic Stress Response of Plants: Antioxidant, Cofactor, and Regulator. *Front Plant Sci* 2021;12:598173.

[67] Wheeler G L, Jones MA, Smirnoff N. The biosynthetic pathway of vitamin C in higher plants. *Nature* 1998;393:365–369.

[68] Smirnoff N, Conklin PL, Loewus FA. Biosynthesis of ascorbic acid in plants: a renaissance. *Annu Rev Plant Physiol Plant Mol Biol* 2001;52:437.

[69] Wolucka BA, van Montagu M. GDP-Mannose3', 5'-epimerase forms GDP-L-gulose, a putative intermediate forthe novo biosynthesis of vitamin C in plants. *J Biol Chem* 2003;278:47483–47490.

[70] Valpuesta V, Botella MA. Biosynthesis of L-ascorbic acid in plants: new pathways for an old antioxidant. *Trends Plant Sci* 2004;9:573–577.

[71] Elkelish A, Qari SH, Mazrou YSA, Abdelaal KAA, Hafez YM, Abu-Elsaoud AM et al. Exogenous ascorbic acid induced chilling tolerance in tomato plants through modulating metabolism, osmolytes, antioxidants, and transcriptional regulation of catalase and heat shock proteins. *Plan. Theory* 2020;9:431.

[72] Mittler R, Vanderauwera S, Gollery M, van Breusegem F. Reactive oxygen gene network of plants. *Trends Plant Sci* 2004;9:490–498.

[73] Noctor G, Mhamdi A, Foyer CH. The roles of reactive oxygen metabolism in drought: not so cut and dried. *Plant Physiol* 2014;164:1636–1648.

[74] Smirnoff N. Ascorbate, tocopherol and carotenoids: metabolism, pathway engineering and functions. In: *Antioxidants and Reactive Oxygen Species in Plants*, ed N. Smirnoff. Oxford: Blackwell publishing, 2005.

[75] Anjum NA, Gill S, Gill R, Hasanuzzaman M, Duarte AC, Tuteja N et al. Metal/metalloid stress tolerance in plants: role of ascorbate, its redox couple, and associated enzymes. *Protoplasma* 2014;251:1265–1283.

[76] Miyaji T, Kuromori T, Takeuchi Y, Yamaji N, Yokosho K, Shimazawa A et al. AtPHT4;4 is a chloroplast-localized ascorbate transporter in Arabidopsis. *Nat Commun* 2015;6:5928–6928.

[77] Vidal-Meireles A, Tóth D, Kovács L, Neupert J, Tóth SZ. Ascorbate deficiency does not limit nonphotochemical quenching in Chlamydomonas reinhardtii. *Plant Physiol* 2020;182:597–611.

[78] Trubitsin BV, Mamedov MD, Semenov AY, Tikhonov AN. Interaction of ascorbate with photosystem I. *Photosynth Res* 2014;122:215–231.

[79] Terzi R, Kalaycıoglu E, Demiralay M, Saglam A, Kadioglu A. Exogenous ascorbic acid mitigates accumulation of abscisic acid, proline and polyamine under osmotic stress in maize leaves. *Acta Physiol Plant* 2015;37: 43.

[80] Piovesana S, Capriotti AL, Cavaliere C, La Barbera G, Montone CM, Chiozzi RZ, Laganà A. Recent trends and analytical challenges in plant bioactive peptide separation, identification and validation. *Anal Bioanal Chem* 2018;410:3425–3444.

[81] Yi-Shen Z, Shuai S, FitzGerald R. Mung bean proteins and peptides: nutritional, functional and bioactive properties. *Food Nutr Res* 2018;62:1290.

[82] Namdeo AG, Ingawale DK. Ashwagandha: Advances in plant biotechnological approaches for propagation and production of bioactive compounds. *J Ethnopharmacol* 2021;271:113709.

[83] Essien SO, Young B, Baroutian S. Recent advances in subcritical water and supercritical carbon dioxide extraction of bioactive

compounds from plant materials. *Trends Food Sci Technol* 2020; 97:156-169.

[84] Ünlü AE. Green and Non-conventional Extraction of Bioactive Compounds from Olive Leaves: Screening of Novel Natural Deep Eutectic Solvents and Investigation of Process Parameters. *Waste Biomass Valori* 2021.

[85] Peanparkdee M, Iwamoto S. Encapsulation for Improving in Vitro Gastrointestinal Digestion of Plant Polyphenols and Their Applications in Food Products. *Food Rev Int 2020.*

Chapter Reviewed by:

Professor Dr Junior Borella, Institute of Biological Sciences, Federal University of Rio Grande - FURG, Rio Grande, Rio Grande do Sul, Brazil.

Dr[a] Simone Ribeiro Lucho, Department of Botany, Biology Institute, Federal University of Pelotas - UFPel, Capão do Leão, Rio Grande do Sul, Brazil.

In: Bioactive Compounds …
Editors: G. Yildiz and G. İzli
ISBN: 978-1-68507-361-9

Chapter 8

POLYSACCHARIDES FROM TROPICAL FRUITS

L. X. Lopez-Martinez[1,*], J. A. Domínguez-Dávila[1], G. A. González-Aguilar[2] and S. A. Enriquez-Valencia[2]

[1]CONACYT-Centro de Investigación en Alimentación y Desarrollo A.C. Sonora, México

[2]Coordinación de Alimentos de Origen Vegetal, Centro de Investigación en Alimentación y Desarrollo, A.C. Sonora, México

ABSTRACT

Tropical fruits grow in tropical or subtropical climates, in the geographical area extending from latitude 30 ° south to latitude 30 ° north. The temperature conditions in this area vary between 16 to 36 ° C during the year. These fruits are important to developing countries from a nutritional and bioactive point of view. Their pulp, peel and

* Corresponding Author's Email: leticia.lopez@ciad.mx.

seeds are rich sources of bioactive compounds such as phenolics acids, polyphenols, betalains, carotenoids, vitamin C and polysaccharides.

Polysaccharides are polymeric carbohydrate molecules composed of long chains of monosaccharides units bounds together by glucosidic bonds that generate various structures that determine different biological properties such as antioxidant, immunomodulatory, antitumor, antiinflammatory, antidiabetic, and antifatigue effects for which they have attracted growing interest to be studied. Hence, this chapter aims to describe the most relevant information on bioactive polysaccharides reported in tropical fruits and their relationship with potential beneficial health effects.

Keywords: bioactive properties, polysaccharides, tropical fruits

Introduction

Tropical fruits are produced by a large and varied group of perennial to herbaceous species with diverse textures, colors, aromas and shapes. They contain many different biologically active compounds, such as phenolic compounds, vitamin C, fiber, carotenoids and polysaccharides [1]. Tropical fruits originate in the warm climate of the tropics, from latitude 30° south to 30° north. In the tropical region, temperatures vary between 16 to 36°C, averaging 25°C throughout the year [2]. In 2018, the Food and Agriculture Organization of the United Nations (FAO) [3] estimated that the world production of the main tropical fruits reached around 100 million tons. India is the world's major tropical fruit-producing country, followed by Brazil, China and Mexico. The most widely consumed tropical fruits that have reached substantial commercial success include avocado (*Persea americana*), papaya (*Carica papaya*). mango (*Mangifera indica*) and pineapple (*Ananas comosus).* They are accessible almost all year in regions with temperate climates [3]; however other fruits like acerola (*Malpighia emarginata*), guava (*Psidium guajava*), bitter gourd (*Momordica charantia*), jackfruit (*Artocarpus heterophyllus*), jujube (*Ziziphus jujuba*), lychee (*Litchi chinensis*), longan (*Dimocarpus longan*), noni (*Morinda citrifolia*), passion fruit (*Passiflora edulis*) and pomegranate (*Punica granatum*) are not as widely available and are therefore difficult to find on the general market, as these fruits are imported from traditional producing countries. Regardless, demand for these fruits is growing in international markets, due in part to their appearance, flavor, high nutritive value, and variety of bioactive compounds [4], among them, polysaccharides.

Polysaccharides are polymeric carbohydrate macromolecules constituted of long chains of monosaccharide units linked by glycosidic bonds [5]; their monomeric composition and structure determine different biological properties [6]. Polysaccharides extracted from tropical fruits have gained interest due to their particular bioactivities, such as antioxidant [7], antifatigue [8], antiinflammatory [9], hepatoprotective [10], immunomodulatory [11], antidiabetic [12] and anticancer [13] which will be discussed in the following sections.

EXTRACTION OF POLYSACCHARIDES FROM TROPICAL FRUITS

Extraction is the first and most important step in the isolation of bioactive polysaccharides. Diverse types of extraction techniques have been used for this purpose, such as hot water extraction (HWE), alkaline water extraction, microwave-assisted extraction, ultrasonic-assisted extraction, ultra-high pressure extraction and enzyme-assisted extraction [14].

The use of HWE has been reported as one of the most common methods; it consists of immersing the raw material into distilled water at high temperatures, followed by filtration, centrifugation, precipitation, concentration, and freeze-drying, steps that ultimately produce a crude polysaccharide extract. Operating conditions like temperature, time, and solvent/material ratio affect the yield of extracted polysaccharides. HWE has been used in the extraction of polysaccharides from tropical fruits, including acerola, bitter gourd, guava, jujube, litchi, noni, passion fruit and pomegranate [15, 16, 17]. The scientific names, common names, and places of origin of tropical fruits discussed in this chapter are shown in Table 1.

Table 1. Origin, scientific and common name of the tropical fruits

Name	Origin	Reference
Acerola (*Malpighia emarginata*)	Brazil	[18]
Bitter gourd (*Momordica charantia*)	Amazon	[19]
Guava (*Psidium guajava*)	Mexico South America	[20]
Jackfruit (*Artocarpus Heterophyllus*)	Malaysia	[21]
Jujube (*Zizyphus jujuba*)	China	[23]
Lychee (*Litchi chinensis*)	Malay Peninsula	[23]
Longan *(Dimocarpus longan*)	China	[24]
Mango (*Mangifera indica*)	Northeast India	[25]
Noni (*Morinda citrifolia*)	Southeast Asia	[26]
Papaya (*Carica papaya*)	South Mexico	[27]
Passion fruit (*Passiflora edulis*)	Brazil	[28]
Pineapple (*Ananas comousus*)	Amazon	[29]
Pomegranate (*Punica granatum*)	Iran	[30]

STRUCTURAL FEATURES

Many different polysaccharides have been isolated from tropical fruits with hot water (HWE) are often additional purified by a combination of procedures such as deproteinization, gel filtration chromatography and ion-exchange chromatography [31]. In particular, ion-exchange chromatography separates acidic and neutral polysaccharides, applying different NaCl solution concentrations as eluent. Gel filtration separates are polysaccharides of various molecular weights, which are subsequently collected, dialyzed, concentrated and lyophilized to produce a pure polysaccharide [6, 32]. The physicochemical and structural characteristics of a polysaccharide principally include its monosaccharide composition, molecular weight, configuration, type and position of its glycosidic bonds [33].

Table 2. Main characterization of polysaccharides from tropical fruits in this chapter

Polysaccharide	Monosaccharide composition	MW (KDa)	Backbone structure	Ref.
Acerola ACSW	UroA:Ara:Gal: Xyl:Rha	61	(1→5)-linked α-Araf	[8]
Guava	GalA:Xyl:Glu: Man:Gal:Rha	ND	(1→5) linkedα- L-Ara (1→3)-linked α- L-Ara	[12]
Jackfruit JFP	Rha:Ara:Gal:Glc: Xyl:GalA	1668	ND	[35]
Jujube HJP3 HJP1	 Rha:Ara:Gal:Glc Rha:Ara:Gal:Glc. GalA	 29.36 67.62	 ND ND	[36]
Longan LPD2	Glu:Gal:Man: Ara	ND	→6)-Glc-(→ Gal-(1 →6)-Man-(1→ →5)-Ara-(1→	[11]
Mango MG1 MG2 MG3	Man:Rha:GluA: GalA:Glu:Gal: Xyl:Ara:Fuc	17.8 32.6 2796	α-glycosidically linked.	[37]
Noni	Fuc:Ara:Gal: Glu:Xyl:Man: Rha:GalA:GluA	135.6	ND	[38]
Passion fruit PFCM	GalA:Ara;Rha: Glu:Man:Gal: Xyl: Rib:Fuc		(1 → 4) linked GalA	[39]
Pine apple PAP-2	Rha:Ara:Xyl:Man: Glc: Gal	1910	→4-α-DManp (1→2,4)-α-D-Manp- (1→	[34]
Pomegranate PSP001	ND	110	B-1-3-Galp	[17]

MW: Molecular weight; ND: Not determined; Fuc: fucose, Rha: rhamnose, Ara: arabinose, Gal: galactose, Glu: glucose, Fru: fructose, GalA: galacturonic acid, Man: mannose, Xyl: xylose, Rib: ribose, GluA: glucuronic acid, UroA: uronic acid; Ref.: refernces.

Polysaccharides with various monosaccharide constituents and chemical structures that are isolated from tropical fruits are then analyzed with a combination of methods, for example, high-performance liquid chromatography (HPLC), gas chromatography-mass spectroscopy (GC–MS), gas chromatography (GC), infrared spectroscopy (IR), nuclear magnetic resonance (NMR), Smith degradation, methylation analysis, acid hydrolysis and periodate oxidation, which allow determining their chemical structure and molecular properties [34]. The structural characterization, including monosaccharide chain, monosaccharide composition and ratio, backbone structure and molecular weight of polysaccharides discussed in this chapter are summarized in Table 2.

BIOLOGICAL ACTIVITIES OF FRUIT POLYSACCHARIDES

Polysaccharides from tropical fruits have been studied for their antioxidant, antidiabetic, antiinflammatory, immunomodulatory, hepatoprotective, anticancer and antifatigue effects, among others (Figure 1). These will be discussed in the present section.

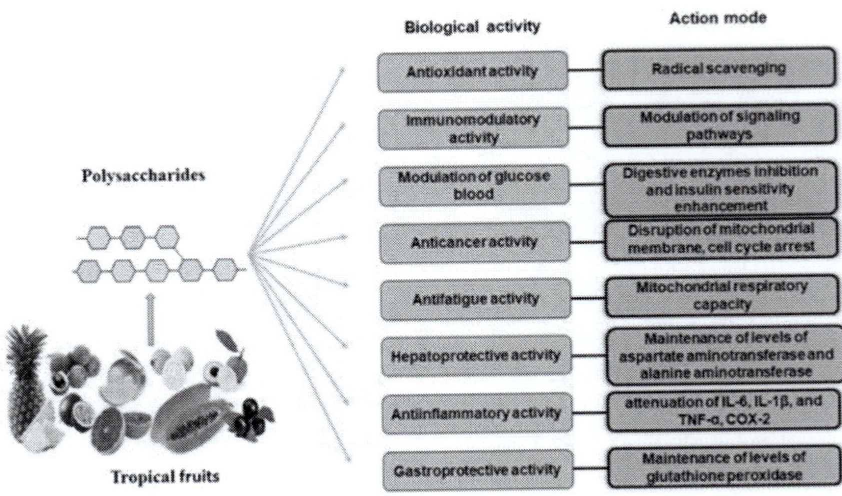

Figure 1. Bioactivities of polysaccharides from tropical fruits.

Extraction is the first and most important step in the isolation of bioactive polysaccharides. Diverse types of extraction techniques have been used for this purpose, such as hot water extraction (HWE), alkaline water extraction, microwave-assisted extraction, ultrasonic-assisted extraction, ultra-high pressure extraction and enzyme-assisted extraction [14].

Antioxidant Activities

Free radicals and oxidative stress cause extensive damage to living organisms and have been associated with various diseases like cancer, arthritis and neurodegenerative diseases, in addition to promoting aging [40, 41]. Polysaccharides from tropical fruits can scavenge free reactive oxygen species (ROS) [35], according to numerous *in vitro* data.

For example, the antioxidant capacity of a polysaccharide (GP70-2) isolated from guava pulp was evaluated according to its DPPH• radical scavenging ability. GP70-2 demonstrates concentration-dependent radical scavenging effects, from 20 to 48%, at a concentration of 2 to 10 mg/mL, with an IC_{50} = 10.8 mg/mL, although its effectiveness was less than that of the control (ascorbic acid) at the same concentrations [42].

In a similar study, Zhang et al. [6] reported values of inhibition of DPPH• and EC_{50} of 97.3% and 0.0672 mg/mL, respectively, for a related water-soluble polysaccharide (GP90). The antioxidant activities of a polysaccharide from jackfruit pulp (JFP) were evaluated by measuring its DPPH• and •OH radical scavenging potential. Results demonstrated that at concentrations of 0.25-4 mg/mL, its scavenging capacities were in the range of 21.82 - 69.64%. At a 2 mg/mL concentration, the maximum scavenging effect of JFP on •OH radicals was 80.10%. Another way through which polysaccharides can exert antioxidant activity is by their stimulating effect on antioxidant enzyme activity, such as SOD), catalase (CAT), and glutathione peroxidase (GSH-Px), which has been previously documented in various models [6, 7]. The *in vivo* antioxidant effects of polysaccharides from tropical fruits have also been investigated. In this sense, Chi et al. [9] studied the effects of oral administration of jujube polysaccharide conjugates (200 mg/Kg) to rats. Results showed increased activities of GSH-Px and SOD of 20.4 and 17.2%, respectively, whereas the level of an important marker of oxidative stress malondialdehyde (MDA) decreased by 31.4%.

Antifatigue Effects

Fatigue is a defensive system in response to life-threatening over-exhaustion and may be associated with muscle soreness, anxiety, and, in extreme cases, depression. If persistent fatigue is not resolved, it may lead to or exacerbate different pathologies, such as increased aging and hypertension and others [43]. The search for natural compounds that reduce fatigue in humans with minimal side effects is ongoing; however, the study of polysaccharides from tropical fruits with antifatigue effects is incipient.

Chi et al. [9] evaluated the oral administration of polysaccharide conjugates from jujube (JJP) (100, 200 and 400 mg/Kg) in rats with chronic fatigue (CF), using the Morris water-maze and open-field tests. Results showed that platform-time and motionless-time in rats treated with JJP were reduced, while crossing-times were increased in a dose-dependent manner compared with those in the rats in CF.

Klosterhoff et al. [8] reported the effects of an oral administration of arabinan-rich pectin from acerola fruit (ACSW), for 28 days, at doses of 50, 100 and 200 mg/kg. They showed that the treatment could lengthen swimming time, where the animals showed increased times of exhaustion (95, 151 and 129 min, respectively) compared to the control (53 min). Furthermore, the mitochondrial respiratory capacity of the skeletal muscle was enhanced at 200 mg/kg.

Antidiabetic Effects

Almost 463 million people were estimated to live with diabetes mellitus (DM) worldwide in 2019, with a projection of 578 million of people by 2030 [44]. Various antidiabetic drugs are currently available on the market; however, they are correlated with undesirable effects [45]. For this reason, the study of phytochemical substances which could act as possible adjuvant treatments has been explored in the latest years.

Delaying glucose absorption by the inhibition of α-amylase and α-glucosidase (carbohydrate hydrolyzing enzymes) has been extensively utilized as a strategy to reduce postprandial hyperglycemia as therapy for type 2 diabetes [46]. In this regard, α-amylase and α-glycosidase inhibition assays are effective approaches for evaluating the hypo-glycemic activities of polysaccharides *in vitro*. For example, a water-soluble polysaccharide (GP-90) isolated from guava pulp showed remarkable α-glucosidase inhibition capacity with an IC_{50} = 2.27 μg/mL, 1379 times lower than the positive control acarbose (IC_{50} = 3.13 mg/mL) [11]. In a similar study, the polysaccharide GP70-3 from guava fruit also exhibited an important α-glucosidase inhibitory activity *in vitro,* with values of IC_{50} = 2.54 μM, which was 1867 times lower than acarbose (IC_{50} = 4.74 mM) [12].

The anti-hyperglycemic activity of polysaccharides from tropical fruits may be attributed to regulating the intestinal absorption of glucose, improving glucose tolerance or metabolism, or preventing the destruction of β-cells [6]. In this sense, Zhao et al. [479] evaluated the effect of intragastric doses of 200 or 400 mg/Kg of jujube polysaccharides (ZSP) on fructose-treated mice with impaired insulin sensitivity, hyperglycemia and hyperinsulinemia. The 400 mg/Kg dose reduced both serum glucose (from 9.62 to 8.42 mM) and insulin (from 72.73 to 44.61 pmol/L). Insulin resistance of mice treated with 200 or 400 mg/Kg was reduced by 25.2 and 31.3%, respectively, as compared to the control group. In a similar study, a water-soluble polysaccharide from bitter gourd (300 mg/Kg) exhibited significant anti-hyperglycemic effects on streptozotocin-induced diabetic mice; furthermore, it was shown to repair injured pancreatic islets and enhance their insulin secretion [48].

Immunomodulatory Effects

The immune response is a complex mechanism involving a variety of cell types that defend the organism against potential pathogens. Perturbation of this well-balanced process results is associated with multiple disorders and chronic diseases, making the control of the immune system a crucial factor in maintaining health [49]. In this subject, polysaccharides from tropical fruits have been explored as immunomodulatory agents. For example, longan pulp polysaccharides were evaluated at 100 and 200 mg/Kg in mice, where both doses showed potent immunomodulatory properties, according to an enhanced of macrophage phagocytosis and concanavalin A-stimulated splenocyte proliferation [50]. In other studies, the immunomodulatory activities of a jujube polysaccharide conjugate (JPC) were studied by Chi et al. [9] in rats with chronic fatigue syndrome (CFS) related to immune cells. Their results showed that the action of JPC enhanced T-cell proliferation and NK activity. The treatment also increased gene expression of interleukin 2 (IL-2) and subsequent protein secretion, in contrast to a decreased IL-10 secretion. In addition, the treatment differently increased the percentage of $CD4^+$ T cells while slightly decreasing the percentage of $CD8^+$ T cells. Rong et al. [11] isolated a polysaccharide (LPD2) from longan pulp, observed that the treatment with 50 µg/mL of LPD2 increased the mouse splenic lymphocytes by 250% and the mouse macrophage phagocytosis by 205.1%, as compared to the control value.

Anticancer Activity

Polysaccharides isolated from tropical fruits have been shown to act on cancerous cells, primarily via apoptosis induction. Some act through

DNA damage, disruption of mitochondrial membrane, cell cycle arrest and nitric oxide production to kill cancerous cells and prevent their metastasis [51, 36].

The anticancer activities of polysaccharides from mango pomace (MG-1, MG-2, MG-3) against human liver (HepG2), breast cancer cell line (MCF-7), human lung epithelial (A549), human cervical (HeLa), human ovarian cancer cell line (A2780), human colon cancer cell line (HCT-116), and human gastric cancer cell line (BGC-823) were investigated [37]. All the polysaccharides exhibited significant inhibitory effects on these cells in a dose-dependent manner at concentration of 25 to 1000 μg/mL. The results showed that MG-2 possessed a higher inhibitory capacity for A2780 cells than MG-1 and MG-3, MG-3 and MG-2showed relatively higher inhibitory rates against HepG2 cells than MG-1, MG-3 showed higher inhibitory rates against HCT-116 and BGC823 cells than MG-2; MG-3 possessed higher inhibitory rates for A549 and MCF-7 cells than MG-1 and MG-2; MG-1 and MG-3 possessed higher inhibitory rates for A549 and MCF-7 cells than MG-1 and MG-2; MG-1. The authors concluded that these polysaccharides can be developed as potential agents to prevent and treat cancer and other diseases.

The antiproliferative effect of jujube cv. Muzao (HJP1 and HJP3) polysaccharides on human HepG2 cells was studied by Wang et al., 2015 [36]. After 96 h of incubation and 5 mg/mL concentration, the inhibition for HJP3 on the growth of the HepG2 cells at 96 h was 53.80%, but not so for HPJ1, which showed inhibition less than 10%. In another study, Huan et al. [16] showed the antiproliferative effects of lychee polysaccharide on the human liver (HepG2) and human cervical (HeLa) cancer cells, which showed a 41.4 and 26.7% of decrease at a dose of 750 μg/mL, respectively, while the growth of human lung epithelial (A549) cells decreased by 27.2% at a 450 μg/mL dose.

The antiproliferative effects of a pomegranate isolated polysaccharide (PSP001) were evaluated *in vitro* against KB (nasopharyngeal carcinoma), K562 (leukemia) and MCF-7 (breast cancer) cells. Significant effects were observed on K562 and MCF-7 cells after 72 h of incubation at a concentration of 0.001 μg/mL and yielding IC_{50} values of 52.8 and 97.21 μg/mL, respectively. PSP001was cytotoxic against KB and MCF-7 cells at 200 μg/mL, while the same effect was apparent at 1000 μg/mL against K562 cells [17].

Passion fruit polysaccharides (PFCM) were dissolved in water and used to treat mice that were transplanted with sarcoma 180 cells into the left hind groin. Oral doses of PFCM (50 and 100 mg/Kg) inhibited tumor growth by 40.5 and 48.7%, respectively. In contrast, intraperitoneal administration (10 and 25 mg/Kg) significantly increased the treatments' effectiveness up to 70.4 and 72.89%, respectively. Based on these findings, the authors concluded that inhibition was associated with the proportion of extensive coagulative necrosis of the tumors that were later extirpated from the animals [39]. Wang et al. [34] isolated the polysaccharide (PAP-2) from pineapple and used as a treatment against MDA-MB-231 breast carcinoma cell line. They showed that HJP3 considerably inhibited the proliferation of the malignant cells while also exerting lower cytotoxicity to healthy cells.

Antiinflammatory Activities

Inflammation is a complex biological reaction of tissues to various potentially detrimental stimuli, such as pathogens, damaged cells or irritants, consisting of both humoral and cellular responses [52]. The continuous presence of inflammatory mediators like tumor necrosis factor-alpha and (TNF-α) IL-1β can upregulate the production of prostaglandin E2 (PGE2) and nitric oxide (NO) by stimulating the activity of cyclooxygenase-2 (COX-2), inducible NO synthase (iNOS) and microsomal PGE synthase-1 (mPGES-1). These enzymes and mediators play a critical role in developing multiple pathologies, including colitis, atherosclerosis, CF syndrome, cardiovascular disorders, neurodegenerative diseases, and some types of cancer [7, 15].

Various studies have supported the antiinflammatory potentials of polysaccharides from tropical fruits. For example, PFPe, a passion fruit polysaccharide, was applied on carrageenan-induced rat paw edema. Carrageenan provoked edema that reached a maximum level after 3 h, while, at the same time, intraperitoneal PFPe at doses of 0.3, 1.0, or 3.0 mg/Kg decreased it by 34.2, 39.5 and 60.6%, respectively. However, only 3 mg/Kg could decrease the inflammatory response caused by histamine (85.3% reduction), serotonin (5-HT) (58% reduction), or PGE2 (62.2% reduction). These results indicate that PFPe administration decreases the inflammatory response by modulating the release or synthesis of serotonin and histamine by reducing neutrophil migration [36].

Yue et al. [53] described that wild jujube polysaccharides (WJPS) alleviated inflammatory bowel disease in mice with 2,4,6-trinitrobenzene sulfonic acid (TNBS)-induced colitis. WJPS significantly decreased the severity of colitis and improved mucosal damage. According to stimuli tests, these effects were associated with decreased inflammatory reactions by attenuating IL-6, IL-1β, and TNF-α.

Batista et al. [54] showed that noni polysaccharides (PLS) exhibited significant antiinflammatory activity in an acetic acid-induced colitis model in rats. The administration of PLS at doses of 3 mg/Kg decreased IL-1β and TNF-α levels in the colon (1.61 and 2.90 pg/mL); it also inhibited colonic expression of COX-2 (0.32, COX-2/p38), as compared to the colitis group (0.66). Based on their findings, the authors argue that PLS exerted significant antiinflammatory effects and protected the colon against injuries, according to a mechanism based on reducing pro-inflammatory cytokines, COX-2 expression and inflammatory cell infiltration.

Gastroprotective Effects

Studies on tropical fruit about the gastroprotective effect of polysaccharides extracted are scarce. In this sense, rats with ethanol-induced gastrointestinal damage were orally fed with polysaccharides from acerola (1 mg/kg). The results revealed that these polysaccharides maintained GSH levels in the gastric mucosa. GSH is a molecule that shows an essential role in maintaining mucosa integrity as one of the main water-soluble antioxidants. GSH may decrease or even be depleted due to the presence of oxidizing agents such as ethanol that promote an increase of free radicals in gastric tissue [55].

Hepatoprotective Activity

Wang et al. [56] reported that the administration of jujube polysaccharides (ZSP) at 200 or 400 mg/Kg body weight to mice was able to dose-dependently prevent carbon tetrachloride (CCl_4)-induced hepatic injury. Authors report that the 400 mg/Kg dose significantly normalized the serum activities of alanine aminotransferase (ALT), aspartate aminotransferase (AST). According to improvements in the enzyme activities of hepatic GSH-Px and SOD, the animals' antioxidant system was also normalized. In a similar study, bitter gourd polysaccharide (BPI) was evaluated in a CCl_4-induced liver injury mice model. The results showed that BPI reduced ALT and AST by 10.6% and 30.7%, respectively [57]. Alias et al. [58] investigated the effect of guava polysaccharides (PG) against paracetamol-induced liver damage in rats. Their results showed that PG doses of 200 and 400 mg/Kg normalized AST and ALT activities that were apparent in the paracetamol group. These studies suggest that some polysaccharides from tropical fruits can effectively protect the liver from hepatotoxic agents by a mechanism that involves mediating free radical-scavenging activities.

In another study, mice fed high-fat diets (HFD) were treated with 200, 400 and 800 mg/Kg of acerola polysaccharides (ACP) to counter diet-induced hepatic lipid deposition. The HFD increased total hepatic triglycerides, but the treatments decreased by 16.1, 17.9 and 44.2%, respectively, which resulted in an attenuated hepatosteatosis. ACP treatments also ameliorated diet-induced hepatic lesions, according to a near-normal hepatocyte morphology with well-preserved cytoplasm. Additional experimentation showed that hepatic damage decreased in response to an inhibition of the sterol regulatory element-binding protein 1c (SREBP1c) pathway [10].

According to the evidence, tropical fruit polysaccharides can be considered potential auxiliary foods to protect the liver from drug- or diet-related injuries.

CORRELATION OF STRUCTURE AND BIOLOGICAL ACTIVITY

It is recognized that the molecular weight of polysaccharides is directly associated with their biological activities [59, 60].

For example, in the case of antioxidant activity, Li et al. [38] described that polysaccharides with high molecular weight displayed higher antioxidant activities due to the high molecular weight allowing the spatial conformation of the polysaccharides to be maintained.

However, Qi et al. [61] reported that the lowest molecular weight polysaccharides had the highest antioxidant activity due to those polysaccharides with low molecular weight would have more reductive hydroxyl group terminals to eliminate and accept free radicals, besides major solubility. The content of uronic acid is considered to be another important indicator reflecting the antioxidant activity of the polysaccharides. Li et al. [41] reported that polysaccharides rich in uronic acids have higher antioxidant activities due to the capacity of uronic acid to alter the physicochemical properties and solubility of the associated polysaccharide conjugates, and therefore affect the activities of polysaccharides [62].

For anticancer activities, the presence of β-(1→3) linkages in the main chain of the polysaccharide are important because it allows the increase in the activities of immunocompetent cells (Wasser, 2002). However, various chemical structures have been reported for the antitumor polysaccharides, including a backbone composed of (1→5)-

linked-Araf [12]. Also, other authors reported that polysaccharides of low molecular weight exhibited stronger antiproliferative activities [36].

Polysaccharides containing sulphate possess strong immunomodulatory capacity, notably reduced when the polysaccharides are desulphated [63, 64]. However, other authors reported that polysaccharides of low molecular weight exhibited stronger immunomodulatory activity and that the molecular weight of polysaccharides might play a more critical role than sulphate content because low molecular weight might increase the capacity of polysaccharides to identify and bind to immune complexes [65].

The aforementioned is necessary to study the structure of the polysaccharides of tropical fruits and their relation to their bioactivities to better understand their mechanisms of action.

Conclusion

Tropical fruits are an excellent source of bioactive compounds, including polysaccharides that have been shown several bioactivities such as antioxidant, antifatigue, antiinflammatory and immunomodulatory, among others. But their mechanisms need to be explored. The biological activities of polysaccharides seem to be associated with their molecular weight and chemical structures, but the structure of tropical polysaccharides is still not well established. Furthermore, structure features vary among the different polysaccharides present in each tropical fruit. Then, the specific structure that possesses certain biological activities should be determined.

Additionally, studies that establish the relationship of structure and biological function are scarce; thus, more studies that establish a relation between structure and biological activity are necessary.

On the other hand, the application of polysaccharides of tropical fruits to induce health effects looks promising and may be an alternative to use as partial substitutes in foods such as baked goods, yogurts, meat, or others to develop new functional food. Scientifically validated health benefits of their consumption could improve the demand, distribution and marketing of tropical fruits; thus, additional work is required to conclusively the stability of the polysaccharides when they are used in the food industry and their benefits when consumed by human.

REFERENCES

[1] González-Aguilar GA, Robles-Sánchez RM, Martínez-Téllez MA, Olivas GI, Alvarez-Parrilla E, De La Rosa, LA. Bioactive compounds in fruits: health benefits and effect of storage conditions. *Stewart Postharvest Review*, (2008). *4*(3), 1-10.

[2] Dembitsky VM, Poovarodom S, Leontowicz H, Leontowicz M, Vearasilp S, Trakhtenberg S, Gorinstein S. The multiple nutrition properties of some exotic fruits : Biological activity and active metabolites. *Food Research International* 2011; 44:1671-1701.

[3] FAO. *El futuro de la alimentación y la agricultura: Vías alternativas hacia el 2050* [*The Future of Food and Agriculture: Alternative Pathways to 2050*]. Versión resumida. (accessed January 2021) 2018. Disponible en: http://www.fao.org/3/CA1553ES/ca1553es.pdf.

[4] Enriquez-Valencia SA, Salazar-Lopez NJ, Robles-Sanchez M, Gonzalez-Aguilar GA, Ayala-Zavala JF, Lopez-Martinez LX. Bioactive properties of exotic tropical fruits and their health benefits. *Archivos Latinoamericanos de Nutricion* 2020; 70:205-214.

[5] de Jesus Raposo MFA, De Morais MB, De Morais RMSC. Marine polysaccharides from algae with potential biomedical applications. *Marine Drugs* 2015; 13:2967-3028.

[6] Ji X, Peng Q, Yuan Y, Shen J, Xie X, Wang M. Isolation, structures and bioactivities of the polysaccharides from jujube fruit (*Ziziphus jujuba* Mill.): A review. *Food Chemistry* 2017; 227:349-357.

[7] Zhang F, Lin L, Xie J. A mini-review of chemical and biological properties of polysaccharides from *Momordica charantia*. *International Journal of Biological Macromolecules* 2016; 92:246-253

[8] Klosterhoff RR, Kanazawa LK, Furlanetto AL, Peixoto JV, Corso CR, Adami ER, Cordeiro LM. Antifatigue activity of an arabinan-rich pectin from acerola (*Malpighia emarginata*). International *Journal of Biological Macromolecules* 2017; 109:1147-1153.

[9] Chi A, Kang C, Zhang Y, Tang L, Guo H, Li H, Zhang K. Immunomodulating and antioxidant effects of polysaccharide conjugates from the fruits of *Ziziphus Jujube* on Chronic fatigue syndrome rats. *Carbohydrate Polymers* 2015; 122:189-196.

[10] Hu Y, Yin F, Liu Z, Xie H, Xu Y, Zhou D, Zhu B. Acerola polysaccharides ameliorate high-fat diet-induced non-alcoholic fatty liver disease through reduction of lipogenesis and improvement of mitochondrial functions in mice. *Food & Function* 2020; 11:1037-1048.

[11] Rong Y, Yang R, Yang Y, Wen Y, Liu S, Li C, Li W. Structural characterization of an active polysaccharide of longan and evaluation of immunological activity. *Carbohydrate polymers* 2019; 213:247-256.

[12] Jiao Y, Hua D, Huang D, Zhang Q, Yan C. Characterization of a new heteropolysaccharide from green guava and its application as an α-glucosidase inhibitor for the treatment of type II diabetes. *Food and Function* 2018; 9:3997-4007.

[13] Wang L, Tang DQ, Kuang Y, Lin FJ, Su Y. Structural characteristics of pineapple pulp polysaccharides and their antitumor cell proliferation activities. *Journal of the Science of Food and Agriculture* 2015; 95:2554-2561.

[14] Mohan K, Muralisankar T, Uthayakumar V, Chandirasekar R, Revathi N, Ganesan AR, Seedevi P. Trends in the extraction, purification, characterisation and biological activities of polysaccharides from tropical and sub-tropical fruits–A comprehensive review. *Carbohydrate Polymers* 2020; 238:116185.

[15] Kamarazaman IS, Abdul RL, Nik Hasan MK, Ahmad MR, Ayob MKM, Rahim N, Zakaria N. 2017. Antioxidant properties of polysaccharide extract from juice and pomace of *Psidium guajava* (pink guava) and its mitigating effect in early stage of Atherogenesis. *Natural Products Chemistry Research* 2017; 5.

[16] Huang F, Zhang R, Yi Y, Tang X, Zhang M, Su D,... Wei Z. Comparison of physicochemical properties and immunomodulatory activity of polysaccharides from fresh and dried litchi pulp. *Molecules* 2014; 19:3909-3925.

[17] Joseph MM, Aravind SR, Varghese S, Mini S, Sreelekha TT. Evaluation of antioxidant, antitumor and immunomodulatory properties of polysaccharide isolated from fruit rind of *Punica granatum*. *Molecular Medicine Reports* 2012; 5:489-496.

[18] Prakash A, Baskaran R. Acerola, an untapped functional superfruit: a review on latest frontiers. *Journal of Food Science and Technology* 2018; 55:3373-3384.

[19] Kumar DS, Sharathnath KV, Yogeswaran P, Harani A, Sudhakar K, Sudha P, Banji D. *A medicinal Potency of Momordica charantia.* 2010.

[20] Lin HC, Lin JY. Characterization of guava (Psidium guajava Linn) seed polysaccharides with an immunomodulatory activity. *International Journal of Biological Macromolecules* 2020; 154:511-520.

[21] Zhang L, Tu ZC, Xie X, Wang H, Wang H, Wang ZX, Lu Y. Jackfruit (*Artocarpus heterophyllus* Lam.) peel: A better source of antioxidants and a-glucosidase inhibitors than pulp, flake and seed, and phytochemical profile by HPLC-QTOF-MS/MS. *Food Chemistry* 2017; 234:303-313.

[22] Zhang L, Zhao J, Feng C, Liu M, Wang J, Hu Y. Genome-wide identification, characterization of the MADS-box gene family in Chinese jujube and their involvement in flower development. *Scientific Reports* 2017; 7:1-13.

[23] Man S, Ma J, Wang C, Li Y, Gao W, Lu F. Chemical composition and hypoglycaemic effect of polyphenol extracts from *Litchi chinensis* seeds. *Journal of Functional Foods* 2016; 22:313-324.

[24] Zhong K, Wang Q. Optimization of ultrasonic extraction of polysaccharides from dried longan pulp using response surface methodology. *Carbohydrate Polymers* 2010; 80:19-25.

[25] Reddy PVR, Gundappa B, Chakravarthy AK. (2018). Pests of mango. In *Pests and their management* (pp. 415-440). Springer, Singapore.

[26] Dussossoy E, Brat P, Bony E, Boudard F, Poucheret P, Mertz C, Michel A. Characterization, anti-oxidative and antiinflammatory effects of Costa Rican noni juice (*Morinda citrifolia* L.). *Journal of Ethnopharmacology* 2011; 133:108-115.

[27] Fuentes G, Santamaría JM. 2014. Papaya (*Carica papaya* L.): origin, domestication, and production. In *Genetics and genomics of papaya* p. 3-15. Springer, New York, NY.

[28] Lim TK. 2012. Passiflora edulis. In *Edible Medicinal and Non-Medicinal Plants,* p. 147-165. Springer, Dordrecht.

[29] Ding P, Syazwani S. Physicochemical quality, antioxidant compounds and activity of MD-2 pineapple fruit at five ripening stages. *International Food Research Journal* 2016;23.

[30] Chandra R, Babu KD, Jadhav VT, Jaime A, Silva TD. Origin, history and domestication of pomegranate. *Fruit, Vegetable and Cereal Science and Biotechnology* 2010; 2:1-6.

[31] Wasser SP. Medicinal mushrooms as a source of antitumor and immunomodulating polysaccharides. *Applied Microbiology and Biotechnology* 2002; 60:258-274.

[32] Liu X, Luo Y, Zha C, Zhou S, Liu L, Zhao L. Rheological properties of polysaccharides from longan (*Dimocarpus longan* Lour.) fruit. *International Journal of Polymer Science* 2015; 2015.

[33] Nie SP, Xie MY. A review on the isolation and structure of tea polysaccharides and their bioactivities. *Food Hydrocolloids* 2011; 25:144-149.

[34] Wang L, Tang DQ, Kuang Y, Lin YF, Su Y. Structural characteristics of pineapple pulp polysaccharides and their antitumor cell proliferation activities. *Journal of the Science of Food and Agriculture* 2015; 95:2554-2561.

[35] Zhu K, Zhang Y, Nie S, Xu F, He S, Gong D,... Tan L. Physicochemical properties and in vitro antioxidant activities of polysaccharide from *Artocarpus heterophyllus* Lam. pulp. *Carbohydrate Polymers* 2017; 155:354-361.

[36] Hung CF, Hsu BY, Chang SC, Chen BH. Antiproliferation of melanoma cells by polysaccharide isolated from *Zizyphus jujuba*. *Nutrition* 2012; 28:98-105.

[37] Hu H, Zhao Q, Pang Z, Xie J, Lin L, Yao Q. Optimization extraction, characterization and anticancer activities of polysaccharides from mango pomace. *International Journal of Biological Macromolecules* 2018; 117:1314-1325.

[38] Li J, Niu D, Zhang Y, Zeng XA. Physicochemical properties, antioxidant and antiproliferative activities of polysaccharides from *Morinda citrifolia* L. (Noni) based on different extraction methods. *International Journal of Biological Macromolecules* 2020; 150:114-121.

[39] Silva DC, Freitas ALP, Barros FCN, Lins KO, Alves APN, Alencar NM, de Paula RC. Polysaccharide isolated from *Passiflora edulis*: Characterization and antitumor properties. *Carbohydrate Polymers* 2012; 87:139-145.

[40] Sousa SG, Oliveira LA, de Aguiar Magalhães D, de Brito TV, Batista JA, Pereira CMC, dos Reis Barbosa AL. Chemical structure and antiinflammatory effect of polysaccharide extracted from *Morinda citrifolia* Linn (Noni). *Carbohydrate Polymers* 2018; 197:515-523.

[41] Halliwell B. Free radicals and antioxidants: updating a personal view. *Nutrition Reviews* 2012; 70:257-265.

[42] Hua D, Zhang D, Huang B, Yi P, Yan C. Structural characterization and DPPH· radical scavenging activity of a

polysaccharide from Guara fruits. *Carbohydrate Polymers* 2014; 103:143-147.

[43] Chaudhuri A, Behan PO. Fatigue in neurological disorders. *The Lancet* 2004; 363:978-988.

[44] Saeedi P, Petersohn I, Salpea P, Malanda B, Karuranga S, Unwin N, IDF Diabetes Atlas Committee. Global and regional diabetes prevalence estimates for 2019 and projections for 2030 and 2045: Results from the International Diabetes Federation Diabetes Atlas. *Diabetes Research and Clinical Practice* 2019; 157:107843.

[45] Wu J, Shi S, Wang H, Wang S. Mechanisms underlying the effect of polysaccharides in the treatment of type 2 diabetes: A review. *Carbohydrate Polymers* 2016; 144:474-494.

[46] Lordan S, Smyth TJ, Soler-Vila A, Stanton C, Ross RP. The α-amylase and α-glucosidase inhibitory effects of Irish seaweed extracts. *Food Chemistry* 2013; 141:2170-2176.

[47] Zhao Y, Yang X, Ren D, Wang D, Xuan Y. Preventive effects of jujube polysaccharides on fructose-induced insulin resistance and dyslipidemia in mice. *Food and Function* 2014; 5:1771-1778.

[48] Yan JK, Wu LX, Qiao ZR, Cai WD, Ma H. Effect of different drying methods on the product quality and bioactive polysaccharides of bitter gourd (*Momordica charantia* L.) slices. *Food Chemistry* 2019; 271:588-596.

[49] Minzanova ST, Mironov VF, Arkhipova DM, Khabibullina AV, Mironova LG, Zakirova YM, Milyukov VA. Biological activity and pharmacological application of pectic polysaccharides: A review. *Polymers* 2018; 10:1407.

[50] Yi Y, Zhang MW, Liao ST, Zhang RF, Deng YY, Wei ZC, Zhang Y. Structural features and immunomodulatory activities of polysaccharides of longan pulp. *Carbohydrate Polymers* 2012; 87:636-643.

[51] Yao WZ, Veeraperumal S, Qiu HM, Chen XQ, Cheong KL. Anticancer effects of *Porphyra haitanensis* polysaccharides on human colon cancer cells via cell cycle arrest and apoptosis without causing adverse effects *in vitro*. *3 Biotech* 2020; 10:1-11.

[52] Ferrero-Miliani L, Nielsen OH, Andersen PS, Girardin SE. Chronic inflammation: importance of NOD2 and NALP3 in interleukin-1β generation. *Clinical and Experimental Immunology* 2007; 147:227-235.

[53] Yue Y, Wu S, Li Z, Li J, Li X, Xiang J, Ding H. Wild jujube polysaccharides protect against experimental inflammatory bowel disease by enabling enhanced intestinal barrier function. *Food and Function* 2015; 6:2568-2577.

[54] Batista JA, de Aguiar Magalhães D, Sousa SG, dos Santos Ferreira J, Pereira CMC, do Nascimento Lima JV,… dos Reis Barbosa AL. Polysaccharides derived from *Morinda citrifolia* Linn reduce inflammatory markers during experimental colitis. *Journal of Ethnopharmacology* 2020; 248:112303.

[55] de Sousa LB, da Costa Gonzaga ML, de Siqueira Oliveira L, Duarte ASG, Silva LMA, de Brito ES, de Sousa PHM. Polysaccharides from acerola, cashew apple, pineapple, mango and passion fruit co-products: Structure, cytotoxicity and gastroprotective effects. *Bioactive Carbohydrates and Dietary Fibre* 2020; 24:100228.

[56] Wang D, Zhao Y, Jiao Y, Yu L, Yang S, Yang X. Antioxidative and hepatoprotective effects of the polysaccharides from *Zizyphus jujube* cv. Shaanbeitanzao. *Carbohydrate Polymers* 2012; 88:453-1459.

[57] Zhou R, Tu NY, Cong M, Kan GS, Ren DM, Chen H M. Purification and hepatoprotective effect of Momordica polysaccharides. *Food Science and Technology* 2013;2.

[58] Alias A, Othman F, Li AR, Kamaruddin A, Yusof R, Hussan F. Supplementation of *Psidium guajava* (Guava) fruit polysaccharide attenuates paracetamol-induced liver injury by enhancing the endogenous antioxidant activity. *Sains Malaysiana* 2015;44: 1129-1136.

[59] Soltani M, Kamyab H, El-Enshasy HA. Molecular weight (Mw) and Monosaccharide composition (MC): Two major factors affecting the therapeutic action of polysaccharides extracted from

Cordyceps sinensis. *Journal of Pure and Applied Microbiology* 2013; 7:1601-1613.

[60] Zong A, Cao H, Wang F. Anticancer polysaccharides from natural resources: A review of recent research. *Carbohydrate polymers* 2012; 90:1395-1410.

[61] Qi H, Zhao T, Zhang Q, Li Z, Zhao Z, Xing R. Antioxidant activity of different molecular weight sulfated polysaccharides from *Ulva pertusa* Kjellm (Chlorophyta). *Journal of Applied Phycology* 2005; 17:527-534.

[62] Yang X, Zhao Y, Lv Y. In vivo macrophage activation and physicochemical property of the different polysaccharide fractions purified from *Angelica sinensis*. *Carbohydrate Polymers* 2008; 71:372-379.

[63] Leiro JM, Castro R, Arranz JA, Lamas J. Immunomodulating activities of acidic sulphated polysaccharides obtained from the seaweed *Ulva rigida C.* Agardh. *International Immunopharmacology* 2007; 7:879-888.

[64] Qiao D, Luo J, Ke C, Sun Y, Ye H, Zeng X. 2010. Immunostimulatory activity of the polysaccharides from *Hyriopsis cumingii*. *International Journal of Biological Macromolecules* 2010; 47:676-680.

[65] Zhang D, Wang C, Wu H, Xie J, Du L, Xia Z, Cai J, Huang Z, Wei D. 2013. Three sulphated polysaccharides isolated from the mucilage of *mud snail, Bullacta exarata philippi*: Characterization and antitumour activity. *Food Chemistry* 2013; 138:306–314.

Chapter Reviewed by:

Dra. Norma Julieta Salazar Lopez, Universidad Autónoma de Baja California, Lic. en Nutrición, Facultad de Medicina de Mexicali. Dr. Humberto Torres Sanginés S/N, Centro Cívico, Mexicali, Baja California 21000, México.

Dra. Juan Buenrostro Figueroa, CIAD, AC. Centro de Investigación en Alimentación y Desarrollo A.C. Av. 4ta Sur No. 3820, C.P. 33089, Delicias, Chihuahu.

In: Bioactive Compounds … ISBN: 978-1-68507-361-9
Editors: G. Yildiz and G. İzli

Chapter 9

DRUGS FROM THE DEEP: MARINE BIOACTIVE NATURAL PRODUCTS AS DRUG CANDIDATES

***Vasantha Subramoniam Pramitha*[1], *Nikhila Leemon*[2], *Chaithra Prasad*[2] *and Parameswara Panicker Sreejith*[2,*]**

[1]Department of Aquatic Biology and Fisheries,
University of Kerala, Thiruvananthapuram
[2]Department of Zoology, University of Kerala,
Thiruvananthapuram

ABSTRACT

Marine natural product discovery has been acknowledged as the most important multidisciplinary endeavor of bioactive substances and drug leads. Marine flora and fauna are taxonomically different with huge productivity and are pharmacologically active novel chemical

* Corresponding Author's Email: p.sreejith@gmail.com, psreejith@keralauniversity.ac.in, pramivs_vzm@yahoo.com.

signatures that account for more than 90% of the total oceanic biomass. Marine organisms continue to amaze us for their capacity to produce chemically unique and biologically important small molecules. Marine products mediate key interactions which has improved human health as medicines and has helped to illuminate many foundational cellular processes. The major effort in marine product discovery is focused on compounds of cosmetics and pharmaceutical interest. Several promising compounds derived from marine sources showed great potential for development as pharmaceuticals. In the recent past, numerous marine natural therapeutic compounds have been isolated, characterized, identified and are under trials for human uses. The marine drugs have a combination of unprecedented chemical structures, biological activities of therapeutic interest for urgent medical needs and new molecular targets that provide the fuel that sustains. In this chapter, we have tried to collect the significance of marine organisms in natural product research and the application of emerging drugs to this field.

Keywords: drug discovery, marine natural products, screening, bacteria, bioprospecting

INTRODUCTION

Natural Products (NPs) are deemed a major origin for bioprospecting and developing new drug leads [1]. Oceans cover 70% of the earth's surface and represent huge diversity of species. Of the 35 animal phyla identified, 34 are found in the marine environment, and many of these are found only in the marine environment [2]. Harsh marine conditions favored the production of molecules bearing unique structures in terms of diversity, structural and functional features [3, 4]. Marine organisms can be considered the most recent source of bioactive natural products. Approximately 28,500 marine natural products (MNPs) have been identified [5], and the biological activities of MNPs are also reported [6]. The marine environment representing half of the biodiversity offers an enormous resource for novel compounds. Bioactive compounds such as sulphated polysaccharides, antiviral substances and halogenated furanones are currently receiving high attention. Marine macroalgae produce a plethora of structurally unique and biologically active secondary metabolites [7].

Bioactivity Profile of MSMs from Marine Macroalgae

During the 40 years of extensive investigations, more than 6000 compounds have been detected from the marine source and bioactivities of these compounds include ichthyotoxicity, phytotoxicity, cytotoxicity, antibiosis, antiviral, insecticidal, antifeedant and pharmacological activities [8]. According to [9], three marine-derived compounds have reached preclinical trials. One of these didemin - B in phase II trials is an anticancer agent. 'Ecteinasidins' - another group of compounds was proved to have potent solid-tumor activity [10]. An analysis of the phyletic distribution of marine secondary metabolites has revealed that the majority (93%) was confined to four groups (macroalgae, coelenterates, echinoderm and sponges), largely a reflection of the abundance and easy collection of these organisms [11]. Chemical defense in the seaweed *Laurencia obsusta* was also reported [12]. *L. obsusta* produced secondary metabolites, which lowered the palatability to herbivores.

Potential Biological Activities: Antibiotic Activities

According to [13], the use of antibiotics increased significantly due to serious infections, and the pathogenic bacteria becoming resistant to drugs necessitated the development of new antibacterial compounds from marine macroalgae. Bioactive metabolites responsible for antibiotic activities in seaweeds are the halogenated compounds such as haloforms, halogenated alkanes and alkenes, alcohols, aldehydes, hydroquinones and ketones [14]. Terpenoids with antibiotic qualities are reported to be halogenated. Compounds such as sterols and heterocyclic and phenolic compounds exhibited antibiotic properties. It is also found that the antibiotic activity of marine organisms *in vivo* is often only achieved at toxic concentrations [14]. The depsipeptides kahalalide A and F from *Bryopsis* sp. were noted for their *in vitro* activity against *Mycobacterium tuberculosis* [15]. A promising antibacterial agent fimbrolide from *Delisea pulchra* has been examined for its effectiveness as an active ingredient in bacterial antifouling agents [16] and as a possible treatment for chronic *Pseudomonas aeruginosa* infection.

Through the GC/MS analysis of the volatile components of *Jania rubens*, 40 compounds were identified, of which *n*-docosine (6.35%), *n*-eicosine (5.77%) and *n*-tetratriacontane (5.58%) act as major components [17]. Studies [18] used to evaluate the antibacterial activity of fresh and dried samples of *Ulva rigida*. Studies also promoted the antibacterial activities of the extracts from 26 algae species prepared by dichloromethane, methanol, and water against five pathogenic fish bacteria. The highest activities were obtained by the dichloromethane extracts and reported that the most active algal species was *Asparagopsis armata*. Studies have shown that the methanol extract of *Cladophora prolifera* had moderate bactericidal activity against *Staphylococcus aureus* and *Vibrio cholera* [19].

According to [20], the extract of *Sargassum tenessimum* has been reported to possess antibiotic effects due to antibiotic substances sarganin-A and sarganin-B. Similarly, a study on different macroalgae collected from the Southern Indian coast exhibited antibiotic activity towards shrimp and fish pathogens [21]. Studies have indicated the susceptibility of *Candida albicans* to several fatty acids and their 1-glycerides and observed that capric acid, 10 carbon saturated fatty acid causes the fastest and most effective killing of all the three strains of *Candida albicans* tested [22].

ANTIVIRAL ACTIVITIES

An important requirement of antiviral polysaccharide is that it must have very low cytotoxic activities and most of the algal polysaccharides, particularly those of *Aghardhiella tenera* and *Nothogenia fastigiata* have this characteristic feature [23]. Sulphated polysaccharides from red algae showed antiviral activities towards viruses responsible for infectious diseases [24] [25]. These polysaccharides were found to be active during the first stage of the RNA virus replication [26].

λ-carrageenan [27] and partially cyclized μ/f-carrageenan from *Gigartina skottsbergii* has potent antiviral effects against different strains of HSV types 1 and 2 during the virus adsorption stage. Carrageenans from cystocarpic and tetrasporophytic stages of *Stenogramme interrupta* showed similar antiherpetic activity [28]. The sulphated polysaccharide of *Schizymenia pacifera* inhibited HIV reverse transcriptase *in vitro* [29] at a later stage in HIV replication. Some agaroids, such as high molecular weight galactan sulphate from *Gracilaria corticata* have antiviral properties against HSV types 1 and 2 [30].

Fucoidan has antiviral properties towards viruses such as RSV, HIV [31], HSV types 1 and 2 and human cytomegalovirus [32]. Fucoidan can inhibit the development of cytopathic effect (CPE) and protect cultural cells from infection caused by viruses. Fucoidans from *Adenocystis utricularis*, *Undaria pinnatifida*, *Stoechospermum marginatum* and *Cystoseira indica* exhibited antiviral activities against HSV-I and HSV-2 without cytotoxicity for Vero cell cultures [33]. The hot water extract of *Hydroclathrus clathratus* contained the main carbohydrate components that exhibited the antiviral activities against various strains of HSV, including the acyclovir-resistant strain and H1-3, a compound fractionated from this showed a dose-dependent anti-HSV activity also [34].

DL-galactan hybrids extracted from the red seaweed *Gymnogongrus torulosus* exhibited *in vitro* antiviral properties, and these compounds were very active against both viruses with inhibitory concentration 50% (IC_{50}) values in the range 0.6-16 μg/ml for HSV-2 and 0.19-1.7 μg/ml for

DEN-2 [35]. Sphingosine compound from *Ulva fasciata* extract also responsible for antiviral activity [36]. The crude water extract, as well as one fraction obtained from *Grateloupia indica* by anion exchange chromatography, had potent anti-HSV activity and their inhibitory concentration values of 0.12-1.06 μg/ml were much lower than its cytotoxic concentration [37]. The hot water extract of *Cystoseira myrica* exhibited antiviral activity against herpes simplex virus type 1 (HSV-1) not only during absorption of the virus to the cells but also on post attachment stages of virus replication [38].

ANTIFUNGAL ACTIVITIES

According to [39], strong antifungal activity was observed in the ethanol extract of *Gelidium amansii,* which strongly inhibited the growth of *F. oxyssporam*. Four diterpenoides isolated from *S. marginatum* which exhibited antifungal activity [40]. *Rhodophyllus membranacea* contained polyhalogenated indols with antifungal activity [41]. A few of the macroalgal extracts inhibited fish pathogenic fungi *Penicillium citrinum* [42].

ANTIPROTOZOAN ACTIVITIES

Crude extract of *Codium iyengarii*, *Ulva reticulata*, *Ulva rigida*, *Scinaia indica*, *Centroceras clavulatum* and *Botryocladia leptopoda* showed good activity against *Leishmania in vitro* [43]. Tetrahydrofucans isolated from Southern Australian brown algae, *Notheia anomala* exhibited antihelmintic property against marine nematodes [44]. The retardation in larval hatching and larval mortality of *Myloidogyne javanica* with the extract of *Sargassum* sp. [46] also reported [45]. The terpenoid [47] compounds in seaweeds might be responsible for larval mortality. Two new diterpenes together with fucoxanthin extracted from *Dictyota* exhibited potent antimalarial activity [48].

Anticoagulant Activities

The anticoagulant activities of marine algal species have recently been reviewed extensively [49]. Anticoagulant activity of sulfated polysaccharides has been identified from brown seaweeds [50] and the green seaweed *Codium cylindricum* [51].

Fucoidans have *in vivo* and *in vitro* heparin-like antithrombic and anticoagulant activities that are mediated by blood coagulation inhibitors such as heparin cofactor II or antithrombin III [52]. According to [53], the anticoagulant activity results from direct fucan-thrombin interaction, and it usually increases with the amount of sulphation. Sulphated fucans from *Fucus vesiculosus* and *Ascophyllum nodosum* have been patented as anticoagulant substances. The anticoagulant activity of *Laminaria angustata* is 200 units/mg compared to heparin (140 units/mg) [54]. The anticoagulant properties of fucoidans were mainly determined by the fucose sulfated chains [55]. The oversulfated fucoidan showed four times higher anticoagulant activity than native fucoidan [56]. However, sulphated arabinogalactan also has been reported [57]. Sulphated polysaccharide with anticoagulant properties was also isolated from the fermented brown seaweed *Sargassum fulvellum* [58].

The hot water extract of *Monostroma nitidum* contained the active anticoagulant fraction [59]. A low molecular weight fucoidan (LMWF) obtained from *Ascophyllum nodosum* by acid hydrolysis has *in vitro* anticoagulant activity, indicating that branched structures were not always necessary for anticoagulant activity [60]. The results of [61], indicated that a 2-sulfated, 3-linked α-L-galactan is a potent thrombin inhibitor mediated by antithrombin or heparin co-factor [62]. In addition, heparin-like polysaccharides possess the capacity to inhibit cancer cell proliferation, angiogenesis, heparanase-mediated cancer cell invasion and cancer cell adhesion to vascular endothelia via adhesion receptors such as selectins. It is reviewed that fucoidan's anticoagulant properties from *Fucus vesiculosus* were determined by thrombin inhibition mediated via plasma antithrombin-III *in vitro* and *in vivo* experiments [63]. Pholorofucofuroeckol A is an antiplasmin inhibitor found in *Ecklonia kurome* [64].

Glucosidase Inhibition

α-D-Glucosidase and β-D-glucosidase are hydrolytic enzymes of α and β-glycosidic linkages of glucosides. These enzymes are involved in several important biological processes such as digestion of carbohydrates into glucose or processing of the oligosaccharides moieties of glycoproteins [65]. Glucosidase inhibitors such as nojirimycin, acarbose, and AO-128 have been obtained from natural and semi-synthetic origins [66]. The extracts of brown algae exhibited stronger glucosidase inhibitory effects than the extracts of red algae. In particular, *Hizikia fusiforme* and *Sargassum thunbergii* were showed strong inhibitory potentials. Inhibitory activities of algal extracts against β-D-glucosidase by their ethyl acetate, n-butanol, and water fractions were also reported [67]. Most of the algal extracts examined and showed inhibitory activities against α-D-glucosidase. Studies revealed that extracts of red algae, *Hypnea musciformis* had higher glucosidase inhibition than the extracts of the green algae, *Ulva fasciata* [68].

Cytotoxicity

Cytotoxic hydroazulene diterpenes is isolated from the brown alga, *Cystoseira myrica* [69]. It is also noted that ulvan has cytotoxicity or cytostaticity targeted to normal or cancerous colonic epithelial cells [70]. The meroterpenes, usneoidone E, and Z from *Cystophora usneoides* reported having antitumour properties [71]. Fucoidan of *Laminaria japonica* can inhibit hepatoma QGY7703 cells. Hexane soluble fractions of the ethanol extract of *Sargassum marginatum* and *Sargassum swartzii* exhibited cytotoxicity towards brine shrimp. Methanol extracts of *Plocamium telfairiae* have induced cytotoxicity and caspase-dependent apoptosis in HT-29 human colon carcinoma cells [72].

Table 1. Some of the activity profiles

Species	Compound	Activity
Fucus vesiculosus Enteromorpha linza	Methanol extract	Antibacterial [78]
Landsburgia quercifolia	1,4,-Naphthoquinone	Cytotoxicity [79]
Undaria pinnatifida	Polysaccharide	Anti-leukemic [80]
Sargassum tortile	Dihydroxysargaquinone	Cytotoxicity [81]
Laminaria angustata	Fucoidan	Anticoagulant [54]
Ecklonia kurome	Fucan sulfate	Anticoagulant [82]
Enteromorpha flexulosa	Sulphono-glycolipid	Antibacterial [57]
Halimeda tuna	Diterpene	Antiviral [83]
Ecklonia kurome	Fucan sulfate	Anticoagulant [84]
Cystoseira usreoides	Meroterpenes	Antitumour [71]
Gracilaria lemaneiformis	Dichloromethane	Haemagglutination [85]
Ecklonia kurome	Fucan sulfate	Antithrombin [86]
Laminaria japonica	Hot water extract	Antimutagenic [87]
Ascophyllum nodosum	Fucans	Antitumour [88]
Dunaliella primolecta	Methanol extract	Antibiotic [89]
Dilophus liguatus	Diterpenoides	Cytotoxic [90]
Gloisiphonia verticilasis	Gloiosiphones	Antimicrobial [91]
Chnoospora fastigata	Organic extract	Antifungal [92]
Sargassum kjellmanianum	DGE	Immunosuppressive [77]
Agardhiella tenera	Sulfated polysaccharide	Anti –HIV [24]
Nothogenia fastigiata	Sulfated polysaccharide	Anti – viral [25]
Tydemania expeditionis	Cycloarthanol sulfates	Protein tyrosine kinase inhibitor [26]
Sargassum thunbergii	Ethyl acetate fraction	α -and β- glucosidase inhibition [67]
Ulva reticulata	Benzene, diethylether	Antibacterial [93]
Laminaria japonica	Fucoidan	Antitumor [34]
Ascophyllum nodosum	Fucoidan	Anticomplementary [94]
Nothogenia fastigata	Xylo-mannans	Anti-viral [95]
Fucus vesiculosus	Fucoidan	Anticoagulant [96]
Nothogenia fastigiata	Sulphated-xylomannan F6	Antiviral [97]
Laminaria japonica	Fucoidan	Anti-viral [98]
Caulerpa taxifolia	Caulerpenyne	Antialgal [99]

Table 1. (Continued)

Species	Compound	Activity
Laurencia obtusa	Sesquiterpene	Antibacterial [100]
Pteroeladia capillacea	Sulfated galactan	Anti -HSV - 1 and 2 [101]
Nothogenia fastigiata	Sulfated xylogalactans	Anti-HSV [102]
Portieria hornemannii	Halomon	Cytotoxic [103]
Delesseria sanguinea	Sulfated xylogalactans	Anticoagulant [104]
Porphyra tenera	Crude extract	Antimutagenic [105]
Hizikia fusiforme	Hot water extract	Release of tumour necrosis factor [106]
Cladophoropsis vaucheriaeformis	Methanol extract	Cytotoxic [107]
Sargassum johnstonii	Lipid fraction	Antibacterial [108]
Laurencia obtusa	5β-Hydroxyaplysistatin	Antifouling [109]
Delisea pulchra	Hexane extract	Antifouling [110]
Hizikia fusiforme	Fucoxanthin	Antioxidant [111]
Gigartina skottsberjii	Carrageenan	Anti – HSV [27]
Laminaria japonica	Fucoidan	Anti-hypertensive [112]
Monostroma nitidum	Aqueous extract	Microalgal enhancer [113]
Ulva lactuca	Steam distillation	Bromoperoxidase [114]
Corallina pilulifera	Crude extract	Algicidal [75]
Eisenia bicyclis	Fucoidan	Anti-tumour [115]
Ulva lactuca	3-o-β-D glucopyranosyl-	Antibacterial [116]
Enteromorpha flexuosa	Unsaponified fractions	Antibacterial [117]
Caulerpa sertularioides	Dicholoromethane fraction	Telomerase inhibition [118]
Dictyota menstrualis	Pachydictyol A	Anti-feedant [119]
Botryocladia occidentalis	Dichloromethane	Anticoagulant [120]
Laminaria japonica	Fucoidans	Antitumor [121]
Jania rubens	16 β-hydroxy-5α-cholestane-3,6-dione	Cytotoxic [122]
Botryocladia occidentalis	Dichloromethane	Anticoagulant [123]
Sphaerococcus coronopifolius	Bromosphaerone	Antibacterial [124]
Laurenciasp.	Ma'iliohydrin	Cytotoxic [125]
Sargassum sp.	Hedols	Cytotoxic [126]
Laurenciasp.	Mailiohydrin	Cytotoxic [127]
Laurencia mariannensis Laurencia nidifica	Halogenated etabolites	Antibacterial [128]
Codium cylindricum	Polysaccharide	Anticoagulant [51]
Codium dwarkense	Sulphated polysaccharide	Anticoagulant [129]
Bostrychia montagnei	Sulfated galactans	Cytotoxic [130]
Ascophyllum nodosum	Fucoidan	Anticoagulant [60]
Laurencia yonaguniensis	Brominated diterpene	Antibacterial [131]
Ulva lactuca	Hot water extract	Immunostimulating [132]
Laurencia obtusa	Palisadin A	Antifouling [133]
Gymnogongrus torulosus	DL-galactan hybrids	Antiviral [35]

Species	Compound	Activity
Codium iyengarii	Steroidal glycosides	Antibacterial [134]
Undaria pinnatifida	Sulphated polyanions	Antiviral [135]
Laminaria japonica	Fucoidans	Antioxidant [136]
Botryocladia occidentalis	Dichloromethane	Anticoagulant [61]
Cissus quandrangularis L.	Ethyl acetate	Antioxidant [137]
Bryopsis plumosa *Petalonia fascia*	Ethanol extract	Antimicrobial [138]
Laurencia majuscula	Elatol, iso-obtusol	Antibacterial [139]
Rhodomela confervoides	Bromophenols	Antibacterial [140]
Laminaria japonica	Fucoidans	Antioxidant [141]
Fucus vesiculosus	Fucoidans	Anticoagulant [63]
Caulerpa prolifera	Acetylene sesquiterpenoid esters	Antibacterial Anti algal [142]
Fucus vesiculosus	Polyhydroxylated fucophlorethol	Antibacterial [143]
Cystoseira myrica	Hydroazulene Diterpenes	Cytotoxic [69]
Cystoseira canariensis	Sulfated polysaccharides	Bind to the myostatin proein [144]
Codium shameelii	Crude extract	Antifungal [145]
Adenocystis utricularis	Fucoidans	Antiviral [32]
Ecklonia kurome	Phlorotannins	Algicidal [74]
Ceramium nitens	Ethanol and lipid soluble extracts	Antibacterial [146]
Stoecheospermum margilatum	Methanol extract	Antibacterial [19]
Sargassum patens	Sulphated polysaccharides	Antiviral [147]
Dictyota pfaffi	Diterpenes	Antiviral [148]
Chondrus crispus	Oxylipins	Produce innate immunity [149]
Chondrus crispus	Oxylipins	Produce innate immunity [149]
Rhodomela confervoides	Bromophenols	Cytotoxic [150]
Dictyota menstrualis	Heterofucans	Anticoagulant [151]
Enteromorpha compressa	Crude extract	Anti-allergic [152]
Laurencia chondrioides	Sesquiterpenoides	Antibacterial [153]
Botryocladia occidentalis	Dichloromethane	Anticoagulant [154]
Cystoseira tamariscifolia	Ethanol extracts	Antibacterial [155]
Caulerpa racemosa	Dichloromethane	Antiherpetic [156]
Undaria pinnatifida	Fucoidans	Antiviral [157]
Sargassum wightii	Crude extract	Immunostimulating [158]
Ulva fasciata	Crude extract	Antibacterial [159]

MICROALGAL GROWTH RESPONSE

Studies have indicated the role in microalgal growth of gibberellins, promoting and regulatory growth hormones isolated from marine macroalgae [73]. It is also indicated that the algicidal property of methanol extract of *Hypnea musciformis* against *Nanochloropsis* sp. led to the proliferation of microalgae [21]. Studies have shown the toxic effect of the brown alga *Ecklonia kurome* on red tide phytoplankton [74]. Macroalgal extracts were also screened for growth inhibition of *Cochlodinium polykrikoids* [75]. Various secondary metabolites like Cyclic depsipeptide, terpenes, polysaccharides (carbohydrates, uronic acids, sulphates), polyhydroxylated lactone, bioglycan, cyclic tripeptide, macrocyclic lactones, etc. have been found to be exerting wide range of immunosuppressing activity [75]. Polyunsaturated fatty acids (PUFAs) are the algicidal compounds isolated from *Ulva fasciata* [76].

Immunosuppressant/Immune Enhancement Activities

Wide ranges of immunosuppressive compounds such as 3,4-Dideoxyglucosone-3-ene (DGE) have been detected in *Sargassum kjellmanianum* [77]. Some of the activity profiles, including the chemical nature of the extracts of marine macroalgae, are listed in Table 1.

Acknowledgments

We thank the Kerala Biotechnology Commission, YIPB program, KSCSTE; HRD scheme, Department of Health Research-Start up Grant, Government of India, and the Plan Fund University of Kerala.

References

[1] El-Demerdash, A; Moriou, C; Martin, MT; Petek, S; Debitus, C; Al-Mourabit, A. Unguiculins, AC. cytotoxic bis-guanidine

alkaloids from the French Polynesian sponge, *Monanchora*. sp. *Natural product research*, 2018, 32(13), 1512–1517.

[2] Newman, DJ; Cragg, GM. Drugs and Drug Candidates from Marine Sources: An Assessment of the Current "State of Play". *Planta Med,* 2016, 82, 775–789.

[3] Altmann, KH. Drugs from the Oceans: Marine Natural Products as Leads for Drug Discovery. *Chimia*, 2017, 71, 646–652.

[4] Romano, G; Costantini, M; Sansone, C; Lauritano, C; Ruocco, N; Ianora, A. Marine microorganisms as a promising and sustainable source of bioactive molecules. *Mar Environ Res*, 2017, 128, 58–69.

[5] Patridge, E; Gareiss, P; Kinch, MS; Hoyer, D. An analysis of FDA-approved drugs: natural products and their derivatives. *Drug Discov Today*, 2016, 21, 204−207.

[6] Blunt, JW; Carroll, AR; Copp, BR; Davis, RA; Keyzers, RA; Prinsep, MR. Marine natural products. *Nat Prod Rep*, 2018, 35, 8–53.

[7] Haefner, B. Drugs from the deep: marine natural products as drug candidates. Drug Discov. *Today*, 2003, 8, 536–544.

[8] Davidson, BS. New dimensions in natural products research: cultured marine microorganism. *Curr Opin Biotechnol*, 1995, 6(3), 284–291.

[9] Rinehart, KL. Pharmaceutical agents from marine sources. In: *Second International Biotechnology Conference*. Programme and Abstracts. IMBC '91. 1991, 87 pp.

[10] Zhang, JY; Fu, LW. Advance of several types of important marine antitumor drugs. *Yao Xue Xue Bao*, 2008, 43(5), 435–42.

[11] Attaway, DH; Zaborsky, OR. *Marine Biotechnology*. Vol. I. Pharmaceutical and bioactive natural products. Plenum Press, New York, 1993, pp. 1–44.

[12] Granado, I; Caballero, P. Chemical defense in the seaweed *Laurencia obtuse* (Hudson) Lamouroux. *Sci Mar Bac*, 1995, 59(1), 31–39.

[13] Smith, DS; Kitts, DD. A competitive enzyme-linked immunoassay for domoic acid determination in human body fluids. *Food Chem Toxicol*, 1994, 32, 1147–1154.

[14] Lincoln, RA; Strupinski, K; Walker, JM. Bioactive compounds from algae. *Life Chem Rep,* 1991, 8, 97–183.

[15] el Sayed, KA; Bartyzel, P; Shen, X; Perry, TL; Zjawiony, JK; Hamann, MT. Marine natural products as antituberculosis agents. *Tetrahedron*, 2000, 56, 949–953.

[16] Kjelleberg, S; Steinberg, P. Surface warfare in the sea. *Microbiology Today*, 2001, 2 8, 134–135.

[17] Karabay-Yavasoglu, NU; Sukatar, A; Ozdemir, G; Horzum, Z. Antimicrobial activity of volatile components and various extracts of the red alga *Jania rubens*. *Phytotherapy Research*, 2006, 21(2), 153–156.

[18] Tuney, I; Cadirci, BH; Unal, D; Sukatar, A. In antimicrobial activities of crude extracts of marine algae from the Coast of izmir (Turkey). *Fres Environ Bull*, 2007, 16, 428–434.

[19] Ely, R; Supriya, T; Naik, CG. Antimicrobial activity of marine organisms collected off the coast of South East India. *Journal of Experimental Marine Biology and Ecology*, 2004, 309(1), 121–127.

[20] Rao, PS; Sreenivasa Rao, P; Karmarkar, SM. Biological investigation of Indian Phaeophyceae. Antimicrobial activity of *Sargassam wightii* Greville. *Seaweed Res Utilin,* 1990, 12 (1 & 2), 577–582.

[21] Selvin, J. "Shrimp Disease Management Using Secondary Metabolites Isolated from Marine Organisms", Ph.D. Dissertation, Manonmaniam Sundaranar University, Tirunelveli, India, 2001, 204 pp.

[22] Bergsson, G; Arnfinnsson, J; Steingrimsson, O; Thormar, H. *In vitro* killing of *Candida albicans* by fatty acids and monoglycerides. *Antimicrobial agents and chemotherapy,* 2001, 45, 3209-12.

[23] De Clercq, E. Chemotherapy of human immunodeficiency virus (HIV) infection: anti-HIV agents targeted at early stages in the virus replicative cycle. *Biomed Pharmacother*, 1996, 50, 207–215.

[24] Witvrouw, M; Este, JA; Mateu, MQ; Reymen, D; Andrei, G; Snoeck, R; Ikeda, S; Pauwels, R; Bianchini, NV; Desmytor, J; De-Clereq, E. Activity of a sulfated polysaccharide extracted from the red seaweed *Agardhiella tenera* against human immunodeficiency virus and other enveloped viruses. *Antiviral Chem Chemother*, 1994, 5(5), 297–303.

[25] Damonte, E; Neyts, J; Pujol, CA; Snoeck, R; Anderi, G; Ikeda, S; Witvrouw, M; Reymen, D; Haines, H; Matulewicz, MC; Cerezo, A; Coto, CS; de-Clereq, E. Antiviral activity of a sulfated polysaccharide from the red seaweed *Nothogenia fastigiata*. *Biochem Pharmacol*, 1994, 47(12), 2187–2192.

[26] De Clercq, E. Current lead natural products for the chemotherapy of human immunodeficiency virus (HIV) infection. *Med Res Rev,* 2000, 20, 323–349.

[27] Carlucci, MJ; Ciancia, M; Matulewicz, MC; Cerezo, AS; Damonte, EB. Antiherpetic activity and mode of action of natural carrageenans of diverse structural types. *Antiviral Res*, 1999, 43(2), 93–102.

[28] Caceres, PJ; Carlucci, MJ; Damonte, EB; Matsuhiro, B; Zuniga, EA. Carrageenans from Chilean samples of *Stenogramme interrupta* (Phyllophoraceae): Structural analysis and biological activity. *Phytochemistry*, 2000, 53, 81-86.

[29] Nakashima, H; Kido, Y; Kobayashi, N; Motoki, Y; Neushul, M; Yamamoto, N. Antiretroviral activity in a marine red alga; reverse transcriptase inhibition by an aqueous extract of *Schizymenia pacifica*. *J Cancer Res Clin Onc,* 1987, 113, 413–416.

[30] Mazumder, S; Ghosal, PK; Pujol, CA; Carlucci, MJ; Damonte, EB; Ray, B. Isolation, chemical investigation and antiviral activity of polysaccharides from *Gracilaria corticata* (Gracilariaceae, Rhodophyta). *Int J Biol Macromol,* 2002, 31, 87–95.

[31] Suguwara, I; Itoh, W; Kimura, S; Mori, S; Shimada, K. Furthur characterization of sulfated homopolysaccharides as anti-HIV agents. *Experientia*, 1989, 45, 996–998.

[32] Ponce, NM; Pujol, CA; Damonte, EB; Flores, ML; Stortz, CA. Fucoidans from the brown seaweed *Adenocystis utricularis* extraction methods, antiviral activity and structural studies. *Carbohydr Res,* 2003, 338(2), 153–165.

[33] Mandal, P; Mateu, CG; Chattopadhyay, K; Pujol, CA; Damonte, EB; Ray, B. Structural features and antiviral activity of sulfated fucans from the brown seaweed *Cystoseira indica*. *Antivir Chem Chemother*, 2007, 18, 153–162.

[34] Wang, H; Ooi, EV; Ang, PO. Antiviral activities of extracts from Hong Kong seaweeds. *J Zhejiang Uni Sci B*, 2008, 9(12), 969–976.

[35] Pujol, CA; Estevez, JM; Carlucci, MJ; Ciancia, M; Cerezo, AS; Damonte, EB. Novel DL-galactan hybrids from the red seaweed *Gymnogongrus torulosus* are potent inhibitors of herpes simplex virus and dengue virus. *Antivir Chem Chemother,* 2002, 13(2), 83–9.

[36] Garg, HS; Sharma, T; Bhakuni, DS; Pramanik, BN; Bose, AK. An antiviral sphingosine derivative from green alga *Ulva fasciata*. *Tetrahydron Lett,* 1992, 33, 1641–1644.

[37] Chattopadhyay, K; Mateu, CG; Mandal, P; Pujol, CA; Damonte, EB; Ray, B. Galactan sulfate of *Grateloupia indica*: Isolation, structural features and antiviral activity. *Phytochem*, 2007, 68(10), 1428–1435.

[38] Zandi, K; Fouladvand, M; Pakdel, P; Sartavi, K. Evaluation of *in vitro* antiviral activity of a brown alga (*Cystoseira myrica*) from the Persian Gulf against herpes simplex virus type 1. *African Journal of Biotechnology*, 2007, 6(22), 2511–2514.

[39] Yi, Z; Yin-shan, C; Hai-sheng, L. Screening for antibacterial and antifungal activities in some marine algae from the Fujian coast of China with three different solvents. *Chinese Journal of Oceanology and Limnology*, 2001, 19(4), 327–331.

[40] Shaikh, W; Shameel, M; Hayee-Memon, A; Usmanghani, K; Bano, S; Ahmad, VU. Isolation and characterization of chemical constituents of *Stoechospermum marginatum* (Dictyotales, Phaeophyta) and their antimicrobial activity. *Pak J Pharm,* 1990, *Sci* 3, 1–9.

[41] Brennan and Ericson. Polyhalogenated indoles from the marine alga *Rhodophyllis membranacea* Harvey. *Tetrahedron Letters*, 1978, 19(19), 1637–1640.

[42] Thampi Raj, S. "Studies on microbial diseases of seahorse reared in captivity and their management", Ph.D. Dissertation, Manonmaniam Sundaranar University, Tirunelveli, India, 2002, 133 p.

[43] Sabina, H; Tasneem, S; Kausar, SY; Choudhary, MI; Aliya, R. Antileishmanial activity in the crude extract of various seaweed from the coast of Karachi, Pakistan. *Pak J Bot,* 2005, 37(1), 163–168.

[44] Capon, RJ; Barrow, RA; Rochfort, S; Jobling, M; Skene, C; Lacey, E; Gill, JH; Friedel, T; Wadsworth, D. Marine nematocides: tetrahydrofurans from a Southern Australian brown algae, *Notheia anomala*. *Tetrahedron*, 1998, 54, 2227–2242.

[45] Ara, J; Ehteshamul-Haque, S; Sultana, V; Ghaffar, A; Qasin, R. Use of *Sargassum* species for the control of *Meloidogyne javanica* in okra. *Nematol Medit,* 1997, 25, 125–128.

[46] Ara, J; Sultana, V; Ehteshamul-Haque, S; Qureshi, SA; Ahmad, VU. Bioactivity of seaweeds against soilborne plant pathogens. *Phytologia*, 1998, 85, 292–299.

[47] Abid, M; Sultana, V; Zaki, MJ; Maqbool, MA. Nematicidal properties of *Stoechospermum marginatum*, a seaweed. *Pak. J Phytopathol*, 1997, 9, 143–147.

[48] Jongaramruong, J; Kongkam, N. Novel diterpenes with cytotoxic, anti-malarial and anti-tuberculosis activities from a brown alga *Dictyota* sp. *Journal of Asian Natural Products Research,* 2007, 9(8), 743–751.

[49] Shanmugam, M; Mody, KH; Oza, RM; Ramavat, BK. Blood anticoagulant activity of a green marine algae *Codium dwarkense* (Codiaceae, Chlorophyta) in relation to its growth stages. *Indian Journal of Marine Sciences,* 2004, 30, 49–52.

[50] Carlucci, MJ; Pujol, CA; Ciancia, M; Noseda, MD; Matulewicz, MC; Damonte, EB; Cerezo, AS. Antiherpetic and anticoagulant properties of carrageenans from the red seaweed *Gigartina skottsbergii* and their cyclized derivatives: correlation between structure and biological activity. *Int J Biol Macromol,* 1997, 20, 97–105.

[51] Matsubara, K; Matsuura, Y; Bacic, A; Liao, ML; Hori, K; Keisuke, M; Miyazawa, K. Anticoagulant properties of a sulfated galactan preparation from a marine green algae, *Codium cylindricum*. *Int J Biol Macromol*, 2001, 28, 395–399.

[52] Matou, S; Helley, D; Chabut, D; Bros, A; Fischer, AM. Effect of fucoidan on fibroblast growth factor-2-induced angiogenesis *in vitro*. *Thromb Res*, 2002, 106, 213–221.

[53] Graufel, V; Kloareg, B; Mabeau, S; Durand, P; Jozefonvicz, J. New natural polysaccharides with potent antithrombic activity: fucans from brown algae. *Biomaterials*, 1989, 10, 363–368.

[54] Kitamura, K; Matsuo, M; Yasui, T. Fucoidan from brown seaweed *Laminaria angustata*. Var. *longissima*. *Agric Biol Chem*, 1991, 55(2), 615–616.

[55] Duarte, M; Cardoso, M; Noseda, M. Structural studies on fucoidans from the brown seaweed *Sargassum stenophyllum*. *Carbohydr Res*, 2001, 333, 281–293.

[56] Qiu, J; Shuanglin, D; Changyun, W. Growth inhibition to three red tide microalgae by extracts of *Ulva pertusa*. *Chinese journal of Oceanology and limnology*, 2006, 24(2), 147–153.

[57] Siddhanta, AK; Ramavat, BK; Chauhan, VD; Achari, B; Dutta, PK; Pakrashi, SC. Sulphonoglycolipid from the green alga *Enteromorpha flexulosa* (wulf.) *J Agric Bot Mar*, 1999, 34, 365–367.

[58] Zoysa, MD; Nikapitiya, C; Jeon, YJ; Jee, Y; Lee, J. Anticoagulant activity of sulfated polysaccharide isolated from fermented brown seaweed *Sargassum fulvellum*. *J Appl Phycol,* 2008, 20, 67–74.

[59] Uehara, T; Takeshita, M; Maeda, M. Studies on anticoagulant active arabinan sulfates from the green alga, *Codium latum*. *Carbohydr Res,* 1992, 235, 309–311.

[60] Chevolot, L; Mulloy, B; Racqueline, JA. A disaccharide repeat unit is the structure in fucoidans from two species of brown algae. *Carbohydr Res,* 2001, 330, 529–535.

[61] Pereira, SM; Melo, RF; Mourao, PA. Is there a correlation between structure and anticoagulant action of sulfated galactans and sulfated fucans? *Glycobiology*, 2002, 12, 573–580.

[62] Borgenstrom, M; Warri, A; Hiilesvuo, K; Kakonen, R; Kakonen, S; Nissinen, L; Pihlavisto, M; Marjamaki, A; Vlodavsky, I; Naggi, A; Torri, G; Casu, B; Veromaa, T; Salmivirta, M; Elenius, K. O-sulfated bacterial polysaccharides with low anticoagulant activity inhibit metastasis. *Semin Thromb Hemost,* 2007, 33(5), 547–56.

[63] Kuznetsova, TA; Besednova, NN; Mamaev, AN; Momot, AP; Shevchenko, NM; Zvyagintseva, TN. Anticoagulant activity of fucoidan from brown *algae Fucus evanescens* of the Okhotsk sea. *Bull Exp Biol Med,* 2003, 136, 471–473.

[64] Fukuyama, Y; Kodama, M; Miura, I; Kinzyo, Z; Mori, H; Nakayama, Y; Takahashi, M. Anti-plasmin inhibitor. VI. Structure of phlorofucofuroeckol A, a novel phlorotannin with both dibenzo-1,4-dioxin and dibenzofuran elements from *Ecklonia kurome* Okamura. *Chem Pharm Bull*, 1990, 38, 133–135.

[65] Collin, M; Karlsson, GB; James, W; Butters, TD; Davis, SJ; Gordon, S; Dwek, RA; Platt, FM. The alpha glucosidase inhibitor *N*-butyldeoxynojirimycin inhibits human immunodeficiency virus entry at the level of post-CD4 binding. *J Vrol,* 1995, 69, 5791–5797.

[66] Horii, S; Fukase, H; Matsuo, T; Kameda, Y; Asano, N; Matsui, K. Synthesis and α -D-glucosidase inhibitory activity of N-substituted

valiolamine derivatives as potential oral antidiabetic agents. *J Med Chem,* 1986, 29, 1038–1046.

[67] Kurihara, H; Sasaki, M; Hatano, M. A new screening method for glucosidase inhibitors and its application to algal extracts. *Fisheries Sci,* 1994, 60(6), 759–761.

[68] Pramitha, VS; Lipton, AP; Thangaraj, M. Evaluation of α and β Glucosidase Inhibitory Properties of Macroalgae Using Intestinal Extract of Marine Snail, *Thais rudolphi* (*Lamarck, 1822*). *Indian Journal of Biotechnology,* 2008, 7, 61–65.

[69] Ayyad, SN; Abdel-Halim, OB; Shier, WT; Hoye, TR. Cytotoxic hydroazulene diterpenes from the brown alga *Cystoseira myrica*. Z. *Naturforsch*, 2003, 58c, 33–38.

[70] Kaeffer, B; Benard, C; Lahaye, M; Blottiere, HM; Cherbut, C. Biological properties of ulvan, a new source of green seaweed sulfated polysaccharides, on cultured normal and cancerous colonic epithelial cells. *Planta Med,* 1999, 65, 527–531.

[71] Urones, JG; Arauja, MEM; Brito-Palma, FMS; Basebe, P; Marcos, IS; Moro, RF; Lithgow, AM; Pineda, J. Meroterpenes from *Cystoseira usreoides*. *Phytochemistry*, 1992, 31(6), 2105–2109.

[72] Kim, JH; Hudson, JB; Huang, AM; Bannistes, K; Jin, H; Choi, TJ; Towers, GHN; Hong, YK; De Wreede, RE. Biological activities of seaweed extracts from British Columbia, Canada and Korea. I. Antiviral activity. *Can J Bot Rev Can Bot Rev Can Bot*, 1997, 75(10), 1656-1660.

[73] Voropayev, VM; Chechin, AP. The growth of marine microalgae in cultures in the presence of gibberlins. *Hydrobiol,* 1983, 17, 57–59.

[74] Nagayama, K; Shibata, T; Fujimoto, K; Honjo, T; Nakamura, T. Algicidal effect of phlorotannins from the brown alga *Ecklonia kurome* on red tide microalgae. *Aquacult*, 2003, 218, 601–611.

[75] Jeong, JH; Jin, HJ; Sohn, CH; Suh, KH; Hong, YK. Algicidal activity of the seaweed *Corallina pilulifera* against red tide microalgae. *Journal of applied phycology*, 2000, 12, 37–43.

[76] Alamsjah, MA; Hirao, S; Ishibashi, F; Oda, T; Fujita, Y. Algicidal activity of polyunsaturated fatty acids derived from *Ulva fasciata* and *U. pertusa* (Ulvaceae, Chlorophyta) on phytoplankton. *J Appl Phycol*, 2007, DOI 10.1007/s10811-007-9257-5.

[77] Kato, F; Mizukoshi, S; Aoyama, Y; Matsuoka, H; Tanaka, H; Nakamura, Tsukamoto, M. Immunosuppressive effects of 3,4-dideoxyglucosone-3-ene, an intermediate in the Mallard reaction. *Journal of agricultural and food chemistry,* 1994, 42(9), 2068–2073.

[78] Lustigmann, B; Brown, C. Antibiotic production by marine algae isolated from the New York/New Jersey Coast. *Bull Environ Contam Toxicol*, 1991, 46, 329–335.

[79] Perry, NB; Blunt, JW; Munro, MHG. A cytotoxic and antifungal 1,4 naphthoquinone and related compounds from a New Zealand brown alga, *Landsburgia quercifolia*. *J Nat Prod*, 1991, 54(4), 978-985.

[80] Furusawa, I; Furusawa, S; Chou, SC. Anti-leukemic activity of viva natural, a dietary seaweed extract, on Rausher murine leukemia in comparison with anti-HIV agents, azidothymidine dextran sulfate and pentosan polysulfate. *Cancer Lett,* 1991, 56(3), 197–205.

[81] Numata, A; Kanbara, S; Takahashi, C; Fujiki, R; Yoneda, M; Fujita, E; Nabeshima, Y. Cytotoxic activity of marine algae and cytotoxic principle of the brown alga *Sargassum tortile*. *Chem Pharm Bull,* 1991, 39, 2129–2131.

[82] Gueven, KC; Oezsoy, Y; Ulutin, ON. Anticoagulant fibrinolytic and antiaggregant activity of carrageenans and alginic acid. *Bot Mar*, 1991, 34(5), 429–432.

[83] Koehn, FE; Gunasekara, SP; Neal, DN; Cross, SS. Halitunal, an unusual diterpene aldehyde from the marine alga *Halimeda tuna*. *Tetrahedron Lett*, 1991, 32, 169–172.

[84] Nishino, T; Nagumo, T; Kiyohara, H. Structure characterization of a new anticoagulant fucan sulfate from the brown seaweed *Ecklonia kurome*. *Carbohydr Res,* 1991, 211, 77–90.

[85] Freitas, ALP; Ainouz, IL; Sampaio, AH. Agglutination of enzyme treated erythrocytes by Brazilian marine algae. *Bot Mar,* 1992, 35, 475–479.

[86] Nishino, T; Nagumo, T. The sulfate-content dependence of the anticoagulant activity of a fucan sulfate from the brown seaweed *Ecklonia kurome*. *Carbohydr Res,* 1992, 214, 193–197.

[87] Okai, Y; Higashi-Okai, K; Nakamura, S. Identification of heterogenous anti-mutagenic activities in the extract of edible brown seaweeds, *Laminaria japonica* (Makonbu) and *Undaria pinnatifida* (Wakame) by the umu gene expression system in *Salmonella typhimurium* (TA 1535/Psk 1002). *Mutat Res*, 1993, 303(2), 63–70.

[88] Ellouali, M; Boisson, VC; Durand, P; Jozefonvicz, J. Antitumour activity of low molecular weight fucans extracted from brown seaweed *Ascophyllum nodosum*. *Anticancer Res,* 1993, 13(6A), 2011–2020.

[89] Chang, T; Ohta, S; Kegami, B; Miyata, H; Kashimoto, T; Kondo, M. Antibiotic substances produced by a marine green alga, *Dunaliella primolecta*. *Bioresour Technol,* 1993, 44(2), 149–153.

[90] Bouaicha, N; Pesando, O; Puel, D; Tringali, C. Cytotoxic diterpenoides from the brown alga *Dilophus liguatus*. *J Nat Prod*, 1993, 56(16), 1747–1752.

[91] Chen, JL; Mughaddam, MF; Gerwick, WH. Gloisiphones A and B, novel metabolites from the red marine alga *Gloisiphonia verticilasis*. *J Nat Prod,* 1993, 55(8), 1205–1210.

[92] Vora, JD; Babu, JM; Sridharan, R; Trivedi, GW; Mathur, HH. Antimicrobial activities of marine algae from the Indian coast. *J Mar Biol Assoc India,* 1994, 36(1-2), 322–325.

[93] Charles, SV; Sivalingam, PM. Antibacterial activity of the tropical Chlorophyta *Ulva reticulata* Forsskal. *Second Asia-Pacific conference on algal biotechnology*, 1994, Singapore.

[94] Blondin, C; Fisher, E; Boisson-Vidal, C; Kazatchkine, M; Jozefonvicz, J. Inhibition of complement activation by natural

sulphated polysaccharides (fucans) from brown seaweed. *Mol Immunol*, 1994, 31, 247–253.

[95] Kolender, AA; Matulewicz, MC; Cerezo, AS. Structural analysis of antiviral sulfated alpha-D-(1,3)-linked mannans. *Carbohydr Res*, 1995, 273(2), 179-85.

[96] Nishino, T; Ura, H; Nagumo, T. The relationship between the sulfate content and the antithrombin activity of an alpha (1 arrow right 2)-fucoidan purified from a commercial fucoidan fraction. *Bot Mar*, 1995, 38(3), 187–193.

[97] Pujol, CA; Coto, CE; Damonte, EB. Determination of the antiviral activity of a naturally occurring sulfated xylomannan under various experimental conditions. *Rev Argent Microbiol*, 1995, 27(2), 91–98.

[98] Li, F; Tian, TC; Shi, YC. Study on antivirus effect of fucoidan *in vitro*. *J N Bethune Univ Med Sci*, 1995, 21, 255–257.

[99] Fischel, JL; Lemee, R; Formento, P; Caldani, C; Moll, JL; Pesando, D; Meinesz, A; Grelier, P; Milano, G. Cell growth inhibitory effects of Caulerpenyne from the marine algae *Caulerpa taxifolia*. *Anticancer Research,* 1995, 15, 2155–2160.

[100] Konig, GM; Wright, AD. Sesquiterpene content of the antibacterial dichloromethane extract of the marine red alga *Laurencia obtusa*. *Planta Med*, 1996, 63, 186–187.

[101] Pujol, CA; Errea, MI; Matulewez, MC; Damonte, EB. Antiherpatic activity of S1, an algal derived sulfated galactan. *Phytother Res,* 1996, 10(5), 410–413.

[102] Damonte, EB; Maeulewicz, MC; Cerezo, AS; Coto, CE. Herpes simplex virus – inhibitory sulfated xylogalactans from the red seaweed *Nothogenia fastigiata*. *Chemotheraphy Basel*, 1996, 42(1), 57–64.

[103] Carte, BK. Biomedical potential of marine natural products. *Bioscience*, 1996, 46, 271–286.

[104] Nabil, S; Cosson, J. Seasonal variations in sterol composition of *Delesseria sanguine* (Ceremiales, Rhodophyta). *Hydrobiologia*, 1996, 326-327, 511–514.

[105] Okai, Y; Higashi Okai, H; Yano, Y; Otani, S. Identification of antimutagenic substances in an extract of edible red alga, *Porphyra tenera* (Asadusa-nori). *Cancer Lett*, 1996, (1-2), 235–240.

[106] Okai, Y; Higashi, OK; Ishizaka, S; Yamashita, U. Enhancing effect of polysaccharide from an edible brown alga, *Hijikia fusiforme* (Hijiki), on release of tumour necrosis factor alpha from macrophages of endotoxin nonresponder C3H/HeJ mice. *Nutr Cancer,* 1997, 27(1), 74–79.

[107] Harada, H; Noro, T; Kamei, Y. Selective antitumor activity *in vitro* from marine algae from Japan coasts. *Biol Pharm Bull,* 1997, 20(5), 541–6.

[108] Rao PPS. Biological investigations of Indian phaeophyceae, 17 Seasonal variation of antibacterial activity of total sterols obtained from frozen samples of *Sargassum johnstonii* Setchell et Gardner. *Seaweed Res Util*, 1998, 20(1-2), 91–95.

[109] Steinberg, P; de Nys, R; Kjelleberg, S. Chemical cues for surface colonization. *J Chem Ecol*, 1998, 28, 1935–1951.

[110] Nys, R; Dworjanyn, SA; Steinberg, PD. A new method for determining surface concentrations of marine natural products on seaweeds. *Marine Ecology Progress Series*, 1998, 162, 79–87.

[111] Yan, X; Chuda, Y; Suzuki, M; Nagata, T. Fucoxanthin as the major antioxidant in *Hijikia fusciformis*, a common edible seaweed. *Biosci Biotechnol Biochem*, 1999, 63(3), 605–607.

[112] Li, ZJ; Xue, CH; Lin, H. The hypolipidemic effects and antioxidative activity of sulfated fucan on the experimental hyperlipidemia in rats. *Acta Nutrim Sin*, 1999, 21, 280–283.

[113] Cho, JY; Jin, HJ; Lim, HJ; Whyte, JNC; Hong, YK. Growth activation of the microalga *Isochrysis galbana* by the aqueous extracts of the seaweed *Monostroma nitidum*. *Journal of Applied Phycology*, 1999, 10, 561–567.

[114] Flodin, C; Helidoniotis, F; Whitfield, FB. Seasonal variation in bromophenol content and bromoperoxidase activity in *Ulva lactuca*. *Phytochemistry*, 1999, 51, 135–138.

[115] Song, JQ; Xu, YT; Zhang, HK. Immunomodulation action of sulfate polysaccharide of *Laminaria japonica* on peritoneal macrophages of mice. *Chin J Immunol*, 2000, 16, 70–70.

[116] Awad, NE. Biologically active steroid from the green alga *Ulva lactuca. Phytotherapy Research*, 2000, 14(8), 641–643.

[117] Arun Kumar, K; Rengasamy, R. Antibacterial activities of seaweed extracts/fractions obtained through a TLC profile against the phytopathogenic bacterium *Xanthomonas oryzae* pv. *Oryzae. Botanica marina*, 2000, 43(5), 417–421.

[118] Kanegawa, K; Harada, H; Myouga, H; Katakura, Y; Shirahata, S; Kamei, Y. Telomerase inhibiting activity *in vitro* from natural resources; marine algae extracts. *Cytotechnol*, 2000, 33, 221–227.

[119] Pereira, RC; Cavalcanti, DN; Teixeira, VL. Effects of secondary metabolites from the tropical Brazilian brown alga *Dictyota menstrualis* on the amphipod *Parhyale hawaiensis*. *Mar Ecol Prog Ser*, 2000, 205, 95–100.

[120] Farias, WR; Valente, AP; Pereira, SM; Mourao, PA. Structure and anticoagulant activity of sulfated galactans. Isolation of a unique sulfated galactan from the red algae *Botryocladia occidentalis* and comparison of its anticoagulant action with that of a sulfated galactans from invertebrates. *J Biol Chem*, 2000, 275, 29299–29307.

[121] Shi, ZY; Guo, YZ; Wang, Z. Pharmacological activity of fucoidan from *Laminaria Japonica*. *J Shanghai Fish Univ*, 2000, 9, 268–271.

[122] Ktari, L; Blond, A; Guyot, M. 16 ß- Hydroxy-5α-cholestane-3, 6-dione, a Novel Cytotoxicoxysterol from the red algae *Jania rubens*. *Bioorganic and medicinal chemistry letters*, 2000, 10, 2563–2565.

[123] Farias, WR; Nazareth, RA; Mourao, PA. Dual effects of sulfated D-galactans from the red algae *Botryocladia occidentalis* preventing thrombosis and inducing platelet aggregation. *J Thromb Haemost*, 2001, 86, 1540–1546.

[124] Etahiri, P; Bultel-Ponce, V; Caux, C; Guyot, M. *J Nat Prod*, 2001, 64, 1024p.

[125] Francisco, MEY; Erickson, KL. *J Nat Prod*, 2001, 64, 790p.

[126] Takada, N; Watanabe, R; Suenaga, K; Yamada, K; Uemura, D. *J. Nat. Prod*, 2001, 64, 653p.

[127] Elena, M; Francisco, Y; Erickson, KL. Mailiohydrin, a cytotoxic chamigrene dibromohydrin from a phillippine *Laurencia* species. *J Nat Prod*, 2001, 64(6), 790–791.

[128] Vairappan, CS; Suzuki, M; Abe, T; Masua, M. Halogenated metabolites with antibacterial activity from the Okinawan *Laurencia* species. *Phytochemistry*, 2001, 58, 517–523.

[129] Shanmugam, M; Mody, KH; Oza, RM; Ramavat, BK. Blood anticoagulant activity of a green marine algae *Codium dwarkense* (Codiaceae, Chlorophyta) in relation to its growth stages. *Indian Journal of Marine Sciences,* 2001, 30, 49-52.

[130] Steven, FR; Oliveira, MB; Bucchi, DF; Noseda, M; Duarte, ME. Cytotoxic effects against HeLa cells of polysaccharides from seaweeds. *J. Submicrosc. Cytol Pathol,* 2001, 33(4), 477–84.

[131] Takahashi, Y; Daitoh, M; Suzuki, M; Abe, T; Masuda, M. Halogenated metabolites from the new Okinawan red alga *Laurencia yonaguniensis*. *J Nat Prod*, 2002, 65, 395–398.

[132] Kaladharan, P; Gireesh, R; Smitha, KS. Cost effective medium for the laboratory culture of live feed micro algae. *Seaweed Res Utiln*, 2002, 24(1), 35–40.

[133] Da Gama, BAP; Pereira, RC; Carvalho, AGV; Coutinho, R; Valentin, YY. The effects of seaweed secondary metabolites on biofouling. *Biofouling*, 2002, 18, 13–20.

[134] Ali, MS; Saleem, M; Yamdagni, R; Ali, MA. Steroid and antibacterial steroidal glycosides from marine green alga *Codium iyengarii* Borgesen. *Nat Prod Lett*, 2002, 16, 407–413.

[135] Cooper, R; Drager, C; Elliot, K; Fitton, JH; Godwin, J; Thompson, K. GFS, a preparation of Tasmanian *Undaria pinnatifida* is associated with healing and inhibition of reactivation of Herpes. *BMC Complement Altern Med*, 2002, 2, 11 p.

[136] Li, DY; Xu, RY; Zhou, WZ; Sheng, XB; Yang, AY; Cheng, JL. Effects of fucoidan extracted from brown seaweed on lipid peroxidation in mice. *Acta Nutrim Sin*, 2002, 24, 389–392.

[137] Chidambara Murthy, KN; Vanitha, A; Mahadeva Swamy, M; Ravishankar, GA. Antioxidant and antimicrobial activity of *Cissus quadrangularis* L. *Journal of Medicinal food*, 2003, 6(2), 99–105.

[138] Magallanes, C; Cordova, C; Orozco, R. Antibacterial activity of ethanolic extracts of marine algae from central coast of Peru. *Rev Peru Biol*, 2003, 10(2), 1–2.

[139] Vairappan, CS. Potent antibacterial activity of halogenated metabolites from Malaysian red algae, *Laurencia majuscule* (Rhodomelaceae, Ceramiales). *Biomol Eng*, 2003, 20, 255–259.

[140] Xu, N; Fan, X; Yan, X; Tseng, CK. Screening of marine algae from China for their antitumour activities. *Journal of Applied Phycology*, 2004, 16, 451–456.

[141] Zhang, QB; Yu, PZ; Zhou, GF; Li, ZE; Xu, ZH. Studies on antioxidant activities of fucoidan from *Laminaria japonica*. *Chin. Trad. Herbal Drugs*, 2003, 34, 824–826.

[142] Smyrniotopoulos, V; Abatis, D; Tziveleka, LA; Tsitsimpikou, C; Roussis, V; Loukis, A; Vagias, C. *J Nat Prod*, 2003, 66, 21p.

[143] Sandsdalen, E; Haug, T; Stensvag, K; Styrvold, OB. The antibacterial effect of a polyhydroxylated fucophlorethol from the marine brown alga, *Fucus vesiculosus*. W*orld Journal of Microbiology & Biotechnology*, 2003, 19, 777–782.

[144] Ramazanov, Z; Jimenez del Rio, M; Ziegenfuss, T. Sulfated polysaccharides of brown seaweed *Cystoseira canariensis* bind to serum myostatin protein. *Acta Physiol Pharmacol Bulg*, 2003, 27(2-3), 101–6.

[145] Rizvi, MA; Shameel, M. Studies on the bioactivity and elementology of marine algae from the coast of Karachi, Pakistan. *Hydrobiologia*, 2003, 151-152(1), 463–469.

[146] Freile-Pelegrin, Y; Morales, JL. Antibacterial activity in marine algae from the coast of Yucatan, Mexico. *Botanica marina*, 2004, 47(2), 140–146.

[147] Zhu, W; Chiu, LC; Ooi, VE; Chan, PK; Ang, PO. Antiviral property and mode of action of a sulphated polysaccharide from *Sargassum patens* against herpes simplex virus type 2. *Int J Antimicrob Agents*, 2004, 24, 279–-283.

[148] Barbosa, JP; Pereira, RC; Abrantes, JL; Cirne dos Santos, CC; Rebello, MA; Frugulhetti, IC; Texeira, VL. *In vitro* antiviral diterpenes from the Brazilian brown alga *Dictyota pfaffi*. *Plant Med*, 2004, 70, 856–860.

[149] Bouarab, K; Adas, F; Gaquerel, E; Kloareg, B; Salaiin, JP; Potin, P. The innate immunity of a marine red alga involves oxylipins from both the eicosanoid and octadecanoid pathways. *Plant Physiology*, 2004, 135, 1838–1848.

[150] Xu, N; Fan, X; Yan, X; Li, X; Niu, R; Tseng, CK. *Phytochemistry*, 2003, 62, 1221 p.

[151] Albuquerqu, IRL; Queiroz, KCS; Alves, LG; Santos, EA; Leite, EL; Rocha, HAO. Heterofucans from *Dictyota menstrualis* have anticoagulant activity. *Braz J Med Biol Res*, 2004, 37, 167–171.

[152] Venkata Raman, B; Rao, DN; Radhakrishnan, TM. *Enteromorpha compressa* (L.) *Greville* an edible green alga as a source of anti-allergic principle (S). *Indian Journal of Clinical Biochemistry*, 2004, 19(1), 105–109.

[153] Bansemir, A; Blume, M; Schroder, S; Lindequist, U. Screening of cultivated seaweeds for antibacterial activity against fish pathogenic bacteria. *Aquaculture*, 2006, 252, 79–84.

[154] Matsubara, K. Recent advances in marine algae anticoagulants. Current Med. Chem. *Cardiovasc. Hematol Agents*, 2004, 2, 13–19.

[155] Zineb, S; Mohamed, L; Mohamed, F; Khadija, FZ. Inhibition of growth and mycotoxins formation in moulds by marine algae *Cystoseira tamariscifolia*. *African Journal of Biotechnology*, 2004, 3(1), 71–75.

[156] Ghosh, P; Adhikari, U; Ghosal, PK; Pujol, CA; Carlucci, MJ; Damonte, EB; Ray, B. *In vitro* antiherpetic activity of sulfated polysaccharide fractions from *Caulerpa racemosa*. *Phytochem*, 2004, 3151–3157.

[157] Lee, JB; Hayashi, K; Hashimoto, M; Nakano, T; Hayashi, T. Novel antiviral fucoidan from sporophyll of *Undaria pinnatifida* (Mekabu). *Chem Pharm Bull*, 2004, 52, 1091–1094.

[158] Felix, S; Herald Robins, P; Rajeev, A. Immune enhancement assessment of dietary incorporated marine alga *Sargassum wightii* (Phaeophyceae/Punctariales) in tiger shrimp *Penaeus monodon* (Crustacia/Penaeidae) through prophenoloxidase (proPO) systems. *Indian Journal of Marine Sciences*, 2004, 33(4), 361–364.

[159] Selvin, J; Lipton, AP. Biopotentials of *Ulva fasciata* and *Hypnea musciformis* collected from the peninsular coast of India. *J Mar Sci Tech*, 2004, 12(1), 1–6.

In: Bioactive Compounds …
Editors: G. Yildiz and G. İzli
ISBN: 978-1-68507-361-9

Chapter 10

Application of Agro-Industrial By-Products in Poultry: Enrichment of Eggs and Meat with Bioactive Compounds

Panagiotis Simitzis [*] ***and Michael Goliomytis***
Laboratory of Animal Breeding and Husbandry,
Department of Animal Science, School of Animal Biosciences,
Agricultural University of Athens, Athens, Greece

Abstract

There is an increasing demand for eggs and chicken meat around the world, due to their content in high quality nutrients, such as vitamins, minerals and proteins at a very low price compared to the other animal products. Nowadays, a continuous raise of production cost is evident even in poultry due to the increase in the prices of soybean and cereal grains. As a result, non-conventional low input feeding

[*] Corresponding Author's Email: pansimitzis@aua.gr.

strategies based on agro-industrial by-products (AIBP) appear as a prerequisite for ensuring poultry efficiency. The application of several AIBP in poultry diets could combine the improvements in growth performance and quality characteristics of eggs and meat with the prevention of hazards that are associated with their disposal into the environment. At the same time, enrichment of eggs and meat with bioactive compounds that are contained in AIBP fortifies human organism against the adverse effects of free radicals and oxidation, while production cost is reduced and farmer profit is increased. AIBP potential utilization in commercial poultry production requires their proper management, identification of their nutritive value and application of appropriate processing methods. Nevertheless, elucidation of the metabolism of AIBP bioactive compounds in the organism of hens and broilers is essential and will give us the opportunity to adjust the optimal AIBP supplementation levels in hens and broilers and establish their regular use.

Keywords: agro-industrial by-products, eggs, meat, nutraceuticals, poultry

INTRODUCTION

Serious concerns have been arisen in recent years against the animal production sector, since the land and resources that are utilized for the production of animal feedstuffs could be otherwise used for the coverage of human food needs. At the same time, prices of the main feedstuffs for poultry diets, i.e., cereal grains and soybean products are continuously increasing. As a consequence, non-conventional low input feeding approaches based on the agro-industrial by-products (AIBP) increasingly gain ground as means of reducing the environmental, social and financial cost of poultry production. Apart from their significant nutritional advantages, AIBP contain bioactive compounds that could further improve composition of poultry products and fortify human health [1, 2]. In detail, they can stimulate digestive enzymes secretion (intestinal maltase or pancreatic alpha-amylase, lipase, trypsin and chymotrypsin) and consequently improve nutrient digestibility. As a result, balance of intestinal microbiota (useful vs pathogen bacteria), morphology of intestinal wall and immune system are appropriately regulated leading to improved growth performance [3].

AIBP are regarded as a costless source of high-added value ingredients and their incorporation into poultry diets mitigates the environmental hazards that are associated with their disposal, since they possess a high organic load, while the sustainable use of these functional additives – nutraceuticals inside food chain is accomplished. AIBP application in poultry diets is recommended due to the fact that is in line with the sustainable farming systems and the market demand for natural clean, eco-green and clean egg and chicken meat products [1, 2]. AIBP could be used either raw or after processing, such as drying or any other advanced techniques. However, the variation in AIBP composition among different regions and the scarce data regarding aspects such as their bioavailability, bioactivity, functional pathways, application (mode, dose etc.) and interactions with other ingredients in bird gastrointestinal tract hinder their regular use [4].

As the global population increases and the living standards improve, the demand for animal protein will be continuously raised. The poultry industry has successfully coped with this increased demand until today by achieving enhanced growth performance and feed efficiency through the utilization of the latest improvements in the field of genetics, health, nutrition and farming. However, the immunocompetence of contemporary fast growing strains is increasingly compromised and the equilibrium among fast growth physiological functions and optimum gut health status is under threat. Intestinal mucosa plays an important role in the absorption of nutrients but also protects interior tissues from pathogens by reducing their adhesion and if its functionality is disrupted, health and performance of birds are negatively affected [5]. The incorporation of agricultural residues and wastes into poultry diets could promote birds' normal physiological functions and fortify their health status, since natural ingredients included in AIBP manipulate intestinal microbiota and facilitate the development of useful bacteria [6] leading to improved performance.

Apple Pomace

Apple (*Malus domestica*) processing generates great quantities of solid residues (20-30% of the original fruits) that consist of seed, pulp and skin. The disposal of these wastes, generally known as apple pomace (AP), constitutes a serious environmental challenge, due to their highly biodegradable nature. As an alternative, AP could be incorporated into poultry diets, although there are differences in their nutrient density as a result of the applied processing technique and the initial composition of the residues [7].

Inclusion of apple peel waste at the level of 50-100 g/kg increased gizzard and small intestine weights, high density lipoprotein cholesterol levels and decreased low density lipoprotein cholesterol and malondialdehyde content and apparent ileal protein digestibility. Although, no effect on growth performance was observed at 50 g/kg, addition of apple peel waste at the level of 100 g/kg decreased weight gain of broilers [8]. Incremental levels of dried AP (12-20%) also deteriorated growth performance and modified antibody titer production, intestinal morphology, total antioxidant capacity and blood parameters of broilers [9]. On the other hand, application of AP at lower levels (3-6%) did not affect growth performance, gut morphometry and histopathology of broilers, although ileum and ceca weight and ileum digesta viscosity was increased and activities of α- glucosidase, α-galactosidase, β-galactosidase, β-glucuronidase and xylase were influenced [10]. Zafar et al. [11] also suggested that incorporating AP instead of corn up to 20% in broiler diets had no negative effect on meat yield, and at the same time increased weight gain and feed intake and improved feed conversion rate were observed.

Dried AP dietary supplementation (10-25%) seemed to significantly improve fertility, hatchability rate, seminal total antioxidant capacity and sperm characteristics in aging breeder roosters. These beneficial effects of AP were attributed to the increased levels of PUFA and MUFA in the sperm plasma membrane [12]. Fiialovych and Kyryliv [13] indicated an improvement in the process of egg laying, hatchability and vitality of goslings after dietary incorporation of AP at the level of 7%. AP dietary addition up to 10% combined with 0.05% multi-enzyme supplement improved blood biochemical parameters, egg laying rate and traits, without any further adverse effect in laying hens [14].

CITRUS PULP

Dried citrus pulp (CP) is derived after extraction of the juice from citrus fruits (*Rutaceae* family) and drying of the residues (approximately 25% of the original fruits). It is consisted of peel (65%), segment pulp (60-65%) and seeds (0-10%) and represents a good source of metabolizable energy, crude protein, fiber (pectin) and calcium. Citrus peel also contains phenolic compounds (i.e., flavonoids) with well-known antioxidant properties. CP is a costless alternative feed that can be used especially during summer when forage availability is scarce in countries around the Mediterranean basin [15], leading to a reduction of feeding cost and an increase in farmers' profit [16].

In broilers, dietary addition of dried *Citrus sinensis* peel (DCSP) at the rate of 1.5-3% did not influence final weight, hot carcass weight and carcass yield. Moreover, DCSP supplementation at the level of 3% during the starter period (1-21 days) resulted in the highest values for breast and pancreas weight and ileum length, while the lowest values for breast and thigh weight during the whole period (1-42 days) were observed in birds fed with the 1.5 and 3% DCSP supplemented diets, respectively [17]. At the same time, DCSP inclusion at the dose of 3% led to a reduction of cholesterol, low density lipoprotein and triglycerides values [18], while a depression of feed intake, body weight gain and an increase of feed conversion rate during the starter and grower period was also observed [17].

In another experiment by the same research team [19], dietary supplementation with DCSP extract at the levels of 1000-1250 ppm did not have a significant effect on carcass weight, breast, thigh and wing yields, relative weights of internal organs (gizzard, liver, pancreas, heart) and intestinal length (ileum, duodenum and jejunum), while DCSP extract supplementation at the level of 1000 ppm during the starter period (1-21 days) led to increased final weight and at the level of 1250 ppm during the whole fattening period (1-42 days) resulted in decreased abdominal fat content. Moreover, orange or lemon peel extract dietary inclusion at the level of 200 mg/g increased serum total protein, decreased serum lactate dehydrogenase and creatine phosphokinase activity, without further effects on intestinal morphology and growth performance of broiler chickens under hot conditions [20]. Application of orange extract (1000-2000 mg/l) in drinking water of broiler chickens stimulated body and improved humoral immune system, as shown by the increased white blood cells concentration and lymphocytes, enhanced immune response to sheep red blood cells and IgG and IgM titres and decreased heterophil percentage and heterophil to lymphocyte (H:L) ratio [21].

Abbasi et al. [22] indicated an increase of feed intake and body weight gain and a reduction in liver and abdominal fat content and serum triglyceride levels after dietary inclusion of DCSP at the level of 0.5-2% as an effect of an improvement of digestive traits and immune response, while feed conversion rate was not influenced. Furthermore, dried sweet orange peel addition at the level of 0.8% decreased serum glucose, cholesterol, low and high density lipoprotein, triglyceride concentration and improved total antioxidant status, without any further effect on feed intake, body weight gain, feed conversion rate and carcass traits in broiler chickens [23]. Plasma cholesterol, triglyceride, and aspartate amino-transferase levels linearly decreased with an increase in the levels of citrus waste (2.5-7.5%) incorporated into the diet, while total protein, albumin, globulin, and blood urea nitrogen content was not affected in broilers [24].

However, utilization of dried lemon pulp into the diets of broilers at the level of 2.5-12% decreased final body weight, daily weight gain and impaired feed conversion rate. High levels of dried lemon pulp supplementation (7.5-12%) further led to decreased jejunal crypt depth and antibody titers against influenza disease virus and sheep red blood cells [25]. Diets containing citrus pulp at similar levels (5-10%) increased small intestine relative length and reduced carcass yield. Although, increased PUFA meat content was found in birds fed with the CP supplemented diets at 10%, reduced daily weight gain and increased feed intake and feed conversion rate was also observed [26]. Agu et al. [27] reached to similar conclusions, since a significant decrease in body weight of broilers was observed, when the level of maize replaced by sweet orange peel was above 20%, while feed intake, body weight gain, feed conversion rate, weights of the most important carcass cuts (thigh, breast, back, neck and shoulder) and internal organs (kidney, liver, heart, spleen and lung) were not affected. On the other hand, CP up to the level of 10% did not have an adverse effect on intestinal morphometry, body weight, carcass traits and meat quality in broilers, although a reduced oxidation rate in chicken meat was evident [28].

CP inclusion (20%) in ostrich diet increased PUFA content and $\omega 6/\omega 3$ ratio and reduced meat cooking loss [29]. Moreover, dried CP dietary supplementation up to the level of 12% did not affect growth performance and carcass yield in geese [30]. In laying hens, CP incorporation at the level of 9% improved egg yolk oxidative stability, but impairment of performance and egg quality was observed [31]. In contrast, Nazok et al. [32] stated that dried CP application at the rate of 12% had no adverse effect on performance and egg quality in the early phase of production in laying hens.

Grape Pomace

Grape by-products (approximately 20% of the original fruit) are derived after its destemming, crushing and pressing. As a result, grape pomace (GP) or marc contains stems, peels and seeds and represents a rich source of polyphenols that possess antimicrobial, antioxidant, antiviral, anti-inflammatory, neuroprotective, cardioprotective, hepato-protective and anti-cancer properties [33]. Recent research has highlighted the importance of by-products from wine processing as plant materials that could be used in animal and particularly broiler production [34]. Specific treatment, such as polyethylene glycol application could partially inactivate GP condensed tannins without compromising health status, growth performance, carcass traits and meat quality characteristics of broilers [35, 36].

GP dietary supplementation at a level of 5-60 g/kg in broilers improved intestinal morphology and decreased serum cholesterol levels and increased the antioxidant capacity of breast muscle and PUFAs content, although feed intake, feed efficiency, growth performance and weights of pancreas, liver, spleen and abdominal fat were not affected [37-40]. GP addition at 20-60 g/kg did not affect antibody titer against sheep red blood cells, growth performance, relative length of different small intestine segments, and jejunal morphology indices in heat-stressed broiler chickens. However, thigh, drumstick, bursa and thymus percentages were increased, whereas that of abdominal fat was reduced. GP inclusion also reduced the plasma cholesterol, LDL and MDA concentration, while GPx and SOD activities were increased [41].

Moreover, diet enrichment with red GP at the levels of 2.5-7.5% did not have an effect on weight gain, blood biochemical parameters, serum biochemistry, carcass traits and meat quality characteristics (apart from an increase in meat redness) in broilers [42]. Increased final body, giblets and breast weight was observed in broilers after GP inclusion at the level of 3% into the diet [43], although no further effect on malondialdehyde levels in breast and thigh meat was found [44]. Dietary incorporation of phenolic compounds originated from GP at the levels of 5-10% in broilers decreased meat susceptibility to lipid oxidation and increased PUFA content, without adverse effects on feed intake, feed efficiency and body weight [45]. An increase in PUFA levels and a decrease in lipid oxidation rate of meat, without further effect on meat pH, cooking loss and lightness was also evident as a result of GP application at 2.5-7% [46]. Increased antioxidant capacity of chicken meat without further side effects on productive parameters was observed after GP inclusion at the level of 20% [47]. On the other hand, GP dietary supplementation at lower doses (2.5-10 g/kg) did not affect meat lightness and yellowness, lipid oxidation levels and bacterial spoilage [48].

Addition of GP at 15-60 g/kg further increased the antioxidant capacity of breast muscle, ileal content and excreta, although crude protein ileal digestibility, pancreas, liver, spleen and abdominal fat weights and growth performance were not influenced [49]. Dietary supplementation of grape seed extract and GP at the level of 7.2 and 60 g/kg, respectively, also manipulated intestinal microflora and gut morphology in broilers by increasing the biodiversity degree of intestinal bacteria [50]. Lichovnikova et al. [51] examined the effect of dietary GP on amino acid digestibility, intestinal microflora and antioxidant activity in the sera and liver. Addition at 1.5% did not induce adverse effects on growth traits, while apparent nutrient digestion, diet metabolisable energy, number of the considered beneficial bacteria *Lactobacillus spp.* in ileum and plasma antioxidant activity were positively affected. Furthermore, GP inclusion up to the level of 10% GP did not adversely affect growth performance and improved the antioxidant and immune responses of broilers [52].

In laying hens, GP addition at the levels of 4 or 6% did not significantly affect feed intake, feed efficiency, live weight and egg production. Moreover, eggshell thickness, eggshell ratio, albumen index, egg-specific gravity, egg yolk index, Haugh unit, yolk color, total protein, total cholesterol and triglyceride levels were not influenced, although egg yolk and plasma malondialdehyde and serum glucose levels were declined. GP inclusion at the dose of 4% improved egg weight and both 4% and 6% supplementation levels enhanced liver weight [53]. GP dietary incorporation (1-3%) at the end of the productive cycle of heat-stressed laying hens can be beneficial for animal health as indicated by the increased serum total antioxidant capacity and glutathione peroxidase and superoxide dismutase activities, and exerts positive effects in their performance and egg quality as assessed by increased antioxidant capacity and decreased lipid peroxidation rate in yolk [54]. Moreover, the dietary inclusion of grape seed extract at the level of 1350 mg/kg could enhance egg production, Haugh unit and antioxidant status in laying hens

[55]. In quail, no effect of GP dietary inclusion (2-6%) on egg production, feed intake and feed conversion rate was observed. Although Haugh unit and eggshell thickness were not affected by GP addition, albumen weight, egg specific gravity and egg weight linearly decreased as GP levels increased [56].

In breeder female hens, a supplementation period of 2 weeks with 2% of grape seed extract is adequate to improve the metabolic and laying parameters, assessed as a reduction in the fat tissue and heavier and more resistant eggs, respectively, through a beneficial modification in their intestinal microbiota by the increase of *Bifidobacteriaceae*, *Lactobacilliaceae* and *Lachnospiraceae* populations [57]. On the other hand, semen quality was degraded in breeder cockerels due to higher oxidative stress and DNA-damaged spermatozoa observed as an effect of grape seed extract [58].

Olive Cake

Olive oil extraction generates considerable amounts of pulp, skin and kernels that are potential pollutants and represents 80% of olive fruit. Olive cake (OC) is a significant source of cellulose, hemicellulose, lignin, fatty acids (oleic, linolenic, linoleic and palmitoleic; 12–22%), minerals and phenolic and its application up to 8-10% into poultry diets could improve the quality of derived products, namely increase in MUFA and decrease in SFA content, without impairing bone mineralization, biochemical parameters, expression of selected antioxidant enzymes, growth performance and carcass traits. At the same time, hazards associated with its disposal into the environment and the necessity for costly waste management programs are prevented [59-63]. As also indicated, OC can be safely fed to broilers up to 10-15% during early growth stage (1–28 d), although the ideal production index and productive performance were observed when 5% OC was supplemented with 0.4 g/kg of *Saccharomyces cerevisiae* yeast [64] or when 10% OC was enriched with 500 FTU/kg bacterial E. coli phytase [65, 66]. At the same time, OC can be fed to broiler chickens during the final growth stage (28-49 d) at 10%, a dose that can be increased to 20% when supplemented with 1 g/kg citric acid without adverse effects on feed conversion and health status of broilers [67]. Moreover, the application of *Bacillus licheniformis* in OC supplemented poultry diets (2-4%) resulted in improved nutrient and fat utilization, growth performance and antioxidative response in broilers [68]. OC incorporation in broiler diets may favor digestibility and absorption capacity, by promoting the growth of certain useful bacteria and controlling the spread of pathogenic bacteria by means of competitive exclusion, as indicated by the increase in villus height and crypt depth [69].

According to Papadomichelakis et al. [70], oxidative stability and oleic (18:1n-9) and total MUFA levels are improved by OC dietary addition. However, OC dose should not exceed 50 g/kg of the diet because it can be detrimental to breast meat color, pH24 and oxidative stability. Dietary supplementation with OC and an enzyme blend at similar levels (50 g/kg) further increased the carcass (breast, leg and neck) and offal weights [71], but also reduced serum triglycerides and cholesterol levels, jejunum weight and length [72]. However, OC dietary inclusion in broiler diets at greater levels (82.5-165 g/kg) also appeared to improve daily weight gain, meat antioxidant status and oxidative stability in another study [73]. OC incorporation into broiler diets at 5.0% further improved breast and thigh sensory attributes and antithrombotic properties [74]. Improvement of total antioxidant capacity and redox status and decreased protein and lipid peroxidation rates in plasma and tissues were also observed in broilers that were dietary supplemented with olive oil mill wastewater [75]. OC utilization in broiler diets at the rate of 750 ppm further improved animal growth as result of its anti-inflammatory properties, as assessed by the downregulation of IL-8 expression and the upregulation of TGF-β4, and Bu-1 in the ileum [76]. Moreover, supplementation of OC up to 1500 ppm did not influence growth performance and alleviated some of the negative impacts of a 15.5 h short-term fasting challenge [77].

As indicated, cholesterol and SFAs levels are reduced and that of MUFA and PUFA are increased in egg yolk after OC dietary inclusion, without adverse effects on productive performance at a level up to 16% [78]. Egg and eggshell weight were increased and blood level of triglycerides was reduced in hens fed OC enriched diets (9%), while no effect on egg production, egg mass, feed conversion ratio and feed intake was observed [79, 80]. Moreover, OC could be applied in the laying hens' diets up to 20% when supplemented with 0.1% citric acid without negative effects on blood metabolites, laying performance and egg quality [81].

In quail, OC dietary supplementation (5-7.5%) decreased serum total cholesterol and low-density lipoprotein cholesterol and breast muscle SFA and PUFA levels, while MUFA content in breast muscle was increased [82]. Furthermore, Abd El-Moneim et al. [83] suggested that OC could be incorporated into quail diets up to 10% leading to an improvement of antioxidant status, immune response and growth performance, without any adverse effects on renal and liver function as well as nutrient digestibility. OC addition at similar levels (5-10%) into the diets of laying quails can improve serum lipid profile and antioxidant status, egg cholesterol content and performance during early laying periods [84].

POMEGRANATE PULP

Pomegranate (*Punica granatum L.*) outer peel and seed pulp (PP) is a residue of the pomegranate juice industry and could serve as a valuable source of fat in poultry diets, since it contains several fatty acids, mainly unsaturated, such as linolenic, oleic, palmitic, stearic and gadoleic [85]. At the same time, PP possesses antioxidant, anti-inflammatory, anti-coccidial and anthelmintic properties [86]. As indicated PP addition into broiler diets (2-4 g/kg) improved growth rate, blood serum metabolites, immunological parameters and meat quality characteristics [87]. Furthermore, PP extract supplementation (50 ml/l) in drinking water improved final body weight, daily weight gain, and feed conversion rate, reduced lipid peroxidation rate and glutathione and catalase content in liver [88].

Rajani et al. [89] indicated that PP dietary inclusion (0.5%) reduced the experimental induced ascites mortality and improved meat oxidative stability, without any effect on growth performance of broilers. Furthermore, PP utilization (0.5-2%) in broiler diets reduced cholesterol and increased protein in breast, lowered lipid oxidation values and SFA and increased MUFA, PUFA and n- 3 fatty acids in breast and thigh meat [90]. Addition of PP at greater doses (2-4%) increased final body weight and average daily weight gain in broilers. At 4%, thigh, spleen and gizzard weights were also increased, while serum aspartate aminotransferase activity was decreased. Breast weight and catalase enzyme activity was increased when PP was included in the diet at the level of 8 g/kg [91]. Fermented PP (1-2%) could further increase daily weight gain, reduce fecal ammonia emission and improve feed efficiency in broilers [92]. Similar results were found by Ahmed et al. [93], since addition of fermented PP (0.5-2%) improved growth performance and meat n-3 fatty acids and decreased meat SFA, thiobarbituric acid reactive substances and cholesterol levels. However, raw or fermented PP (5-10 g/kg) can have detrimental effects on the ileum morphology in broilers as indicated by decreased villus height to crypt depth ratio. Malondialdehyde levels in breast meat and *Clostridium perfringens* count in caecum were decreased, while body weight and feed conversion ratio, serum glutathione peroxidase, superoxide dismutase and catalase levels were not affected [94].

Dietary inclusion of PP extract at the level of 1% increased body weight, weight gain, feed intake, carcass weight and dressing percentage and decreased intestine bacterial count, while liver, gizzard and heart weights were not affected [95]. Addition of PP extract into broiler diets at the doses of 0.5-1 g/kg improved carcass yield and meat phenol and flavonoid concentration, while decreased body fat levels, serum total cholesterol and triacylglycerol content [96]. Saleh et al. [97, 98] also found increased levels of n-3 long chain PUFA and improved antioxidant capacity in breast meat after PP extract dietary supplementation at the doses of 0.2-0.3 g/kg. Furthermore, adding an extract that included PP at a rate of 2 ml/l in drinking water improved total antiradical activity and cecal microbial ecology by increasing abundance of lactic acid bacteria, while no effect on growth performance was observed [99].

PP extract at a rate up to 650 mg/kg also improved body weight gain, feed conversion rate, plasma high-density lipoprotein content and meat oxidative stability and decreased levels of plasma triglycerides, alanine aminotransferase, alkaline phosphatase cholesterol and low-density lipoprotein in broilers reared under heat stress, while relative weights of carcass, liver, pancreas, gallbladder, heart, gizzard, bursa of Fabricius, spleen, abdominal fat, thigh and breast were not influenced [100]. At the same time, PP supplementation at 7-10% appeared to improve performance, blood cholesterol and antioxidant status in heat stressed broilers [101]. Urea treated PP (30-50g/kg) further improved body weight gain, feed conversion rate, breast yield, blood glucose, high density lipoprotein and globulin levels, villus height and crypt depth and breast meat oxidative stability, while abdominal fat, plasma albumin, alkaline phosphatase, lactate dehydrogenase, alanine aminotransferase, cholesterol, low-density lipoprotein and malondialdehyde content were decreased in broilers under heat stress [102]. Decrease of total cholesterol levels was also indicated after pomegranate seed oil dietary inclusion at 1.5%, although liver enzyme activities and lipid contents were not affected [103]. At the same time, abdominal fat level, PUFA content and CLA deposition in breast was increased [104]. Moreover, the partial replacement of soybean oil by pomegranate seed oil (40%) resulted in increased levels of punicic and rumenic fatty acids, without any other effect on dressing percentage, carcass traits and breast and thigh muscle physico-chemical composition [105].

In laying hens, PP addition at the rate of 2-4% lowered plasma cholesterol, triglyceride and malondialdehyde content and increased total antioxidant blood capacity and the levels of the antioxidative enzymes (SOD, CAT, and GPx) [106]. Moreover, inclusion of pomegranate seed oil in the diets of laying hens improved laying rate, color and concentration of punicic acid and CLA in yolk [107]. Similar results were indicated in quail, since PP incorporation (2.5-7.5%) improved feed conversion rate, egg production, egg weight, egg mass and villus height to crypt depth ratio, while serum cholesterol, triglyceride and glucose levels were decreased [108]. Increased liver glutathione peroxidase and catalase activities, plasma uric acid level and decreased meat malondialdehyde values were also observed in heat stressed quails dietary supplemented with PP extract at 200 mg/kg [109]. However, PP extract did not influence shedding, viability and morphometry of *Eimeira spp.* and it is therefore not suggested for the prevention/treatment of coccidiosis in quail [110].

Sugar Beet Pulp

Sugar beet pulp (SBP) (40-60% of the original fruit) is an abundant fibrous by-product generated from the sugar refining industry. It contains cellulose, hemicelluloses and pectin and possesses significant antioxidant activity [111]. In broilers, SBP dietary inclusion at 2.5% improved body weight, weight gain, feed conversion rate, performance index and giblet relative weight [112]. Furthermore, addition of SBP at 30g/kg improved broilers' growth from 1 to 10 days of age, but this improvement was not evident afterwards. Relative weight of gastrointestinal tract and gizzard, gizzard digesta content and total tract digestibility were also enhanced [113]. Moderate SBP incorporation (up to 50g/kg) improved nutrient digestibility, feed conversion ratio [114] and gastrointestinal tract development [115] in young birds but higher levels (75g/kg) might impair intestinal mucosa structure and growth performance. At the same time, SBP addition at greater levels (7.5%) decreased body weight, weight gain, low-density lipoprotein and total cholesterol serum levels [116] and villus height and villus height to crypt depth ratio [117]. Moreover, in birds fed SBP supplemented diets (23-92 g/kg) a decreased ileal digestibility of organic matter, crude fat and crude protein was observed, while total serum cholesterol concentration was also reduced with increasing levels of SBP [118]. Moreover, SBP at the level of 100-300 mg/kg demonstrated good anticoccidial activity as assessed by lesion score, oocyst score and oocysts per gram of feces [119].

In laying hens, SBP dietary inclusion (3-7%) increased egg production, weight and mass, improved feed conversion ratio, Haugh unit and yolk color score and decreased egg yolk cholesterol, triglyceride and malondialdehyde levels. SBP also resulted in decreased content of serum alanine aminotransferase, creatinine, aspartate aminotransferase, cholesterol and total lipids [120]. On the other hand, SBP dietary supplementation in quail at 20-40 g/kg did not affect productive performance, egg weight mass, internal and external quality and nitrogen balance but reproductive parameters such as fertility and hatchability rates and nutrient digestibility were deteriorated at 40 g/kg [121].

Tomato Pomace

Tomato pomace (TP) is a residue (around 4% of original weight) of the tomato canning industry that includes seeds, skin and a small amount of pulp. TP contains beneficial nutritional compounds, such as fibers (60%), sugars (25%), proteins (20%), pectins (8%), fat (6%), minerals (4%) and antioxidants (e.g., lycopene) [122, 123]. TP can be incorporated into broiler diets up to 15-20% leading to an improvement of feed economic efficiency without adverse effects on growth performance, feed intake, liver, pancreas, gizzard, abdominal fat and breast weight [124-126]. Furthermore, TP dietary supplementation at 5% alleviated the deleterious effects of heat stress on bone composition, immune response, oxidative status and the activities of serum enzymes, without affecting growth performance parameters in broilers [127]. Juice of tomato waste could be also used in broiler rations up to 120ml/day resulting in an increase of thymus, jejunum, duodenum, caecum and liver relative weight, with no effect on blood profile, growth performance and ileal and caecal microbial populations [128].

Dried TP can be also applied in laying hens' diets up to 10% without detrimental effects on growth performance, daily feed intake, egg weight, egg quality indices, such as shell thickness, shell weight, yolk color and Haugh units, and yolk and serum cholesterol levels [129-132]. In general, as TP concentration enhances from 5 to 10 g/kg, a linear increase in egg production, feed intake, egg weight and yolk color score is evident. At the same time, concentrations of serum and egg yolk beta-carotene, lutein, lycopene and vitamin A increased in TP supplemented groups, whereas malondialdehyde content decreased, indicating decreased yolk lipid peroxidation rates [133, 134]. Salajegheh et al. [135] suggested that dried TP can be applied into the diets of laying hens up to 190 g/kg leading to an increased yolk score, without negative implications on feed intake, body weight, feed conversion rate, egg production, egg weight and mass, eggshell weight and thickness, Haugh unit and total serum protein, albumin, glucose, triglyceride, HDL, LDL and cholesterol levels.

Dried TP can be used in quail diets up to 4-6% without deleterious effects on growth performance [136]. Nikolakakis et al. [137] further suggested that TP dietary inclusion at 5-10% did not affect daily feed intake, feed conversion rate, final body weight, carcass weight, yield, composition and fatty acid profile. On the other hand, Botsoglou et al. [138] observed an increase of PUFA and of PUFA/SFA ratio in the meat of quails fed with a dried TP supplemented diet (10%). However, effects on the oxidative stability of meat were not evident at this level but at 5%. Sahin et al. [139] reached to similar conclusions; malondialdehyde levels in liver, serum and muscle of heat stressed quails decreased after TP dietary incorporation at the levels of 2.5-5%, while an increase in feed intake and live weight gain was observed.

Conclusion

Poultry production is nowadays one of the fastest growing animal production subsectors and the demand for chicken meat and eggs is rapidly increasing worldwide. However, a continuous increase of poultry diet cost is evident due to the high prices of soybean products and cereal grains. At the same time, serious concerns that have been arisen in recent years against the livestock production sector as a result of the utilization of land and resources for animal feed that could be otherwise used for human food production seem to be rational and corrective measures should be implemented. Non-conventional inexpensive nutritional strategies are therefore indispensable, based on alternative feedstuffs, such as the AIBP. As it was demonstrated by the reviewing of the existing literature, growth performance and sensory and nutritional properties of the derived products are usually improved after the inclusion of several AIBP in poultry diets. In general, suggested levels of AIBP incorporation in poultry diets may be 10% for apple and tomato pomace, 9% for olive cake, 3% for citrus pulp, grape pomace and sugar beet pulp and 2% for pomegranate pulp. However, factors such as processing method, harvesting period and geographical location modify the composition of the AIBP derived from each fruit. Moreover, large gaps exist in our knowledge concerning their metabolism and digestion. The understanding of these mechanisms would be beneficial for commercial poultry production, giving us the opportunity to adjust the optimal AIBP inclusion levels for broilers, hens and quails and establish their regular use.

REFERENCES

[1] Azizi M, Seidavi AR, Ragni M, Laudadio V, Tufarelli, V. Practical applications of agricultural wastes in poultry feeding in Mediterranean and Middle East regions. Part 1: Citrus, grape, pomegranate and apple wastes. *Worlds Poult Sci J* 2018;74: 489-98.

[2] Seidavi AR, Azizi M, Ragni M, Laudadio V, Tufarelli V. Practical applications of agricultural wastes in poultry feeding in Mediterranean and Middle East regions. Part 2: Tomato, olive, date, sunflower wastes. *Worlds Poult Sci J* 2018;74: 443-52.

[3] Sugiharto S. Role of nutraceuticals in gut health and growth performance of poultry. *J Saudi Soc Agric Sci* 2016;15: 99–111.

[4] Galanakis CM. Recovery of high added-value components from food wastes: Conventional, emerging technologies and commercialized applications. *Trends Food Sci Technol* 2012;26: 68-87.

[5] Rinttila T, Apajalahti J. Intestinal microbiota and metabolites – implications for broiler chicken health and performance. *J Appl Poult Res* 2013;22: 647–58.

[6] Filazi A, Yurdakok-Dikmen B. Nutraceuticals in Poultry Health and Disease. In: Gupta RC, Srivastava A, Laii R (Eds.). *Nutraceuticals in Veterinary Medicine* (pp. 661-672). Springer, Cham; 2019.

[7] Pieszka M, Gogol P, Pietras M, Pieszka M. Valuable components of dried pomaces of chokeberry, black currant, strawberry, apple and carrot as a source of natural antioxidants and nutraceuticals in the animal diet. *Ann Anim Sci* 2015;15: 475–91.

[8] Heidarisafar Z, Sadeghi G, Karimi A, Azizi O. Apple peel waste as a natural antioxidant for heat-stressed broiler chickens. *Trop Anim Health Prod* 2016;48: 831-35.

[9] Aghili AH, Toghyani M, Tabeidian SA. Effect of incremental levels of apple pomace and multi enzyme on performance, immune response, gut development and blood biochemical parameters of broiler chickens. *Int J Recycl Org Waste Agric* 2019;8: 321-34.

[10] Colombino E, Ferrocino I, Biasato I, Cocolin LS, Prieto-Botella D, Zduńczyk Z, et al. Dried fruit pomace inclusion in poultry diet: growth performance, intestinal morphology and physiology. *J Anim Sci Biotechnol* 2020;11: 63.

[11] Zafar F, Idrees M, Ahmed Z. Use of apple by-products in poultry rations of broiler chicks in Karachi. *Pak J Physiol* 2005;1.

[12] Akhlaghi A, Ahangari YJ, Zhandi M, Peebles ED. Reproductive performance, semen quality, and fatty acid profile of spermatozoa in senescent broiler breeder roosters as enhanced by the long-term feeding of dried apple pomace. *Anim Reprod Sci* 2014;147: 64-73.

[13] Fiialovych L, Kyryliv I. Laying performance, egg quality and hatching results in geese fed with dry apple pomaces. *Acta Sci Pol Zootech* 2017;15: 71-82.

[14] Ghaemi H, Nobakht A, Razzaghzadeh S. The effect of apple pulp and multi enzyme on performance and blood parameters in native laying hens. *J Farm Anim Physiol Nutr* 2014;9: 10–21.

[15] Bampidis VA, Robinson PH. Citrus by-products as ruminant feeds: A review. *Anim Feed Sci Technol* 2006;128: 175-217.

[16] Seidavi A, Zaker-Esteghamati H, Salem AZ. A review on practical applications of *Citrus sinensis* by-products and waste in poultry feeding. *Agrofor Syst* 2020;94: 1581-9.

[17] Ebrahimi A, Qotbi AAA, Seidavi A. The effects of different levels of dried *Citrus sinensis* peel on broiler carcass quality. *Acta Sci Vet* 2013;41: 1169.

[18] Ebrahimi A, Qotbi AAA, Seidavi, A, Edens FW, Laudadio V, Tufarelli, V. Selected plasma constituents of broiler chickens fed different levels of dried sweet orange (*Citrus sinensis*) peels. *J Anim Plant Sci* 2016;26: 949-55.

[19] Ebrahimi A, Qotbi AAA, Seidavi A, Bahar B. The effects of dietary supplementation of *Citrus sinensis* peel extract on production and quality parameters of broiler chicken. *J Appl Anim Res* 2014;42: 445-50.

[20] Akbarian A, Golian A, Gilani A, Kermanshahi H, Zhaleh S, Akhavan A, et al.. Effect of feeding citrus peel extracts on growth performance, serum components, and intestinal morphology of broilers exposed to high ambient temperature during the finisher phase. *Livest Sci* 2013;157: 490-7.

[21] Pourhossein Z, Qotbi AAA, Seidavi A, Laudadio V, Centoducati G, Tufarelli V. (2015). Effect of different levels of dietary sweet orange (*Citrus sinensis*) peel extract on humoral immune system responses in broiler chickens. *Anim Sci J* 2015;86: 105-10.

[22] Abbasi H, Seidavi A, Liu W, Asadpour L. Investigation on the effect of different levels of dried sweet orange (*Citrus sinensis*) pulp on performance, carcass characteristics and physiological and

biochemical parameters in broiler chicken. *Saudi J Biol Sci* 2015;22: 139-46.

[23] Alzawqari MH, Al-Baddany AA, Al-Baadani HH, Alhidary IA, Khan RU, Aqil GM, et al. Effect of feeding dried sweet orange (*Citrus sinensis*) peel and lemon grass (*Cymbopogon citratus*) leaves on growth performance, carcass traits, serum metabolites and antioxidant status in broiler during the finisher phase. *Environ Sci Pollut Res* 2016;23: 17077-82.

[24] Behera DP, Sethi APS, Singh C, Singh U, Wadhwa M. Effect of citrus waste on blood parameters of broiler birds with and without cocktail of enzymes. *Vet World* 2019;12: 483-8.

[25] Basir R, Toghyani M. (2017). Effect of dietary graded levels of dried lemon (*Citrus aurantifulia*) pulp on performance, intestinal morphology, and humoral immunity in broiler chickens. *Int J Recycl Org Waste Agric* 2017;6: 125-32.

[26] Mourão JL, Pinheiro VM, Prates JAM, Bessa RJB, Ferreira LMA, Fontes CMGA, et al. Effect of dietary dehydrated pasture and citrus pulp on the performance and meat quality of broiler chickens. *Poult Sci* 2008;87: 733-43. https://doi.org/10.3382/ps.2007-00411.

[27] Agu PN, Oluremi OIA, Tuleun CD. Nutritional evaluation of sweet orange (*Citrus sinensis*) fruit peel as feed resource in broiler production. *Int J Poult Sci* 2010;9: 684-8.

[28] Diaz-Vargas M, Murakami AE, Pintro PTM, Ospina-Rojas IC, de Souza CHP, Eyng C. Dehydrated citrus pulp in broiler diets. *Can J Anim Sci* 2018;99: 33-40.

[29] Lanza M, Fasone V, Galofaro V, Barbagallo D, Bella M, Pennisi P. Citrus pulp as an ingredient in ostrich diet: effects on meat quality. *Meat Sci* 2004;68: 269–75.

[30] Wang C, Gao GL, Huang JX, Zhang KS, Zhong H, Wang HW, et al. Nutritive value of dry citrus pulp and its effect on performance in geese from 35 to 70 days of age. *J Appl Poult Res* 2017;26: 253-59.

[31] Goliomytis M, Kostaki A, Avgoulas G, Lantzouraki DZ, Siapi E, Zoumpoulakis P, et al. Dietary supplementation with orange pulp (*Citrus sinensis*) improves egg yolk oxidative stability in laying hens. *Anim Feed Sci Technol* 2018;244: 28-35.

[32] Nazok A, Rezaei M, Sayyahzadeh, H. Effect of different levels of dried citrus pulp on performance, egg quality, and blood parameters of laying hens in early phase of production. *Trop Anim Health Prod* 2010;42: 737–42.

[33] Georgiev V, Ananga A, Tsolova V. Recent Advances and Uses of Grape Flavonoids as Nutraceuticals. *Nutrients* 2014;6: 391-415.

[34] Brenes A, Viveros A, Chamorro S, Arija I. Use of polyphenol-rich grape by-products in monogastric nutrition. A review. *Anim Feed Sci Technol* 2016;211: 1-17.

[35] Kumanda C, Mlambo V, Mnisi CM. Valorization of red grape pomace waste using polyethylene glycol and fibrolytic enzymes: physiological and meat quality responses in broilers. *Animals* 2019;9: 779.

[36] Van Niekerk RF, Mnisi CM, Mlambo V. Polyethylene glycol inactivates red grape pomace condensed tannins for broiler chickens. *Br Poult Sci* 2020;61: 566-73.

[37] Goñí I, Brenes A, Centeno C, Viveros A, Saura-Calixto F, Rebole A, et al. Effect of dietary grape pomace and vitamin E on growth performance, nutrient digestibility, and susceptibility to meat lipid oxidation in chickens. *Poult Sci* 2007;86: 508-16.

[38] Aditya S, Ohh SJ, Ahammed M, Lohakare J. Supplementation of grape pomace (*Vitis vinifera*) in broiler diets and its effect on growth performance, apparent total tract digestibility of nutrients, blood profile, and meat quality. *Anim Nutr* 2018;4: 210-4.

[39] Turcu RP, Panaite TD, Untea AE, Șoica C, Iuga M, Mironeasa S. Effects of supplementing grape pomace to broilers fed polyunsaturated fatty acids enriched diets on meat quality. *Animals* 2020;10: 947.

[40] Gungor E, Altop A, Erener G. Effect of Raw and Fermented Grape Pomace on the Growth Performance, Antioxidant Status, Intestinal

Morphology, and Selected Bacterial Species in Broiler Chicks. *Animals* 2021;11: 364.

[41] Hosseini-Vashan SJ, Safdari-Rostamabad M, Piray AH, Sarir H. The growth performance, plasma biochemistry indices, immune system, antioxidant status, and intestinal morphology of heat-stressed broiler chickens fed grape (*Vitis vinifera*) pomace. *Anim Feed Sci Technol* 2020;259: 114343.

[42] Kumanda C, Mlambo V, Mnisi CM. From landfills to the dinner table: Red grape pomace waste as a nutraceutical for broiler chickens. *Sustainability* 2019;11: 1931.

[43] Haščík P, Čech M, Čuboň J, Bobko M, Arpášová H, Pavelková A, et al. Effect of grape pomace supplementation on meat performance of broiler chicken Ross 308. *J Microbiol Biotechnol Food Sci* 2020;10 140-4. https://doi.org/10.15414/jmbfs.2020.10.1.140-144.

[44] Jurčaga L, Bobko M, Haščík P, Bobková A, Demianová A, Belej Ľ, et al. Effect of dietary red grape pomace on lipid oxidation in meat of broiler chickens. *J Microbiol Biotechnol Food Sci* 21;10: e3769.

[45] Chamorro S, Viveros A, Rebole A, Rica BD, Arija I, Brenes A. Influence of dietary enzyme addition on plyphenol utilization and meat lipid oxidation of chicks fed grape pomace. *Food Res Int* 2015;73: 197-203.

[46] Bennato F, Di Luca A, Martino C, Ianni A, Marone E, Grotta L, et al. Influence of grape pomace intake on nutritional value, lipid oxidation and volatile profile of poultry meat. *Foods* 2020;9: 508.

[47] Reyes P, Urquiaga I, Echeverría G, Durán E, Morales MS, Valenzuela C. Wine grape pomace flour in broiler diets effects growth and some meat characteristics. *Anim Prod Sci* 2020;60: 1210-6.

[48] Kasapidou E, Sossidou EN, Zdragas A, Papadaki C, Vafeas G, Mitlianga P. Effect of grape pomace supplementation on broiler meat quality characteristics. *Eur Poult Sci* 2016;80: 1-12.

[49] Brenes A, Viveros A, Goni I, Centeno C, Sayago-Ayerdy SG, Arija I, et al. Effect of Grape Pomace Concentrate and Vitamin E on Digestibility of Polyphenols and Antioxidant Activity in Chickens. *Poult Sci* 2008;87: 307-16.

[50] Viveros A, Chamorro S, Pizarro M, Arija I, Centeno C, Brenes A. (2011). Effects of dietary polyphenol-rich grape products on intestinal microflora and gut morphology in broiler chicks. *Poult Sci* 2011;90: 566-78.

[51] Lichovnikova M, Kalhotka L, Adam V, Klejdus B, Anderle V. (2015). The effects of red grape pomace inclusion in grower diet on amino acid digestibility, intestinal microflora, and sera and liver antioxidant activity in broilers. *Turkish J Vet Anim Sci* 2015;39: 406-12.

[52] Ebrahimzadeh SK, Navidshad B, Farhoomand P, Aghjehgheshlagh FM. Effects of grape pomace and vitamin E on performance, antioxidant status, immune response, gut morphology and histopathological responses in broiler chickens. *S Afr J Anim Sci* 2018;48: 324-36.

[53] Kara K, Guclu BK, Baytok E, Senturk M. (2016). Effects of grape pomace supplementation to laying hen diet on performance, egg quality, egg lipid peroxidation and some biochemical parameters. *J Appl Anim Res* 2016;44: 303-10.

[54] Reis JH, Gebert RR, Barreta M, Boiago MM, Souza CF, Baldissera MD, et al. Addition of grape pomace flour in the diet on laying hens in heat stress: Impacts on health and performance as well as the fatty acid profile and total antioxidant capacity in the egg. *J Therm. Biol* 2019;80: 141-9.

[55] Kaya A, Yildirim BA, Kaya H, Gül M, Çelebi S. The effects of diets supplemented with crushed and extracted grape seed on performance, egg quality parameters, yolk peroxidation and serum traits in laying hens. *Eur Poult Sci* 2014;78: 1-10.

[56] Fróes HG, Jácome IMTD, Tavares RA, Garcia RG, Domingues CHF, Bevilaqua TMS, et al. Grape (*Vitis vinifera*) pomace flour as pigment agent of quail eggs. *Braz J Poult Sci* 2018;20: 183-8.

[57] Grandhaye J, Douard V, Rodriguez-Mateos A, Xu Y, Cheok A, Riva A, et al. Microbiota changes due to grape seed extract diet improved intestinal homeostasis and decreased fatness in parental broiler hens. *Microorganisms* 2020;8: 1141.

[58] Grandhaye J, Lecompte F, Chartrin P, Leconte M, Riva A, Barbe A, et al. Maternal dietary supplementation with grape seed extract in reproductive hens increases fertility in females but decreases semen quality in males of the F1 generation. *PloS one* 2021;16: e0246750.

[59] Sayehban P, Seidavi A, Dadashbeiki M, Ghorbani A, Araújo WAG, Albino LFT. Effects of different levels of two types of olive pulp with or without exogenous enzyme supplementation on broiler performance and economic parameters. *Braz J Poult Sci* 2016;18: 489-500.

[60] Sateri S, Seidavi A, Bouyeh M, Kutzler M, Neumann P, Laudadio V, et al. Effect of olive meal and supplemental enzymes on performance traits, blood biochemistry, humoral immunity response and caecal microbiota of broilers. *S Afr J Anim Sci* 2017;47: 804-12.

[61] Pappas AC, Tsiplakou E, Papadomichelakis G, Mitsiopoulou C, Sotirakoglou K, Mpekelis V, et al. Effects of olive pulp addition to broiler diets on performance, selected biochemical parameters and antioxidant enzymes. *J Hellenic Vet Med Soc* 2019;70: 1687-96.

[62] Pečjak M, Levart A, Salobir J, Rezar V. Effect of the supplementation of olive leaves and olive cake on growth performance and bone mineralisation of broiler chickens. *Acta Fytotech Zootech* 2020;23: 105-11.

[63] de Oliveira CO, Roll AAP, Medeiros Gonçalves FM, Lopes DCN, Xavier EG. Olive pomace for the feeding of commercial poultry: effects on performance, meat and eggs quality, haematological parameters, microbiota and immunity. *Worlds Poult Sci J* 2021; in press.

[64] Al-Harthi MA. The efficacy of using olive cake as a by-product in broiler feeding with or without yeast. *Ital J Anim Sci* 2016;15: 512-20.

[65] Al-Harthi MA. The effect of olive cake, with or without enzymes supplementation, on growth performance, carcass characteristics, lymphoid organs and lipid metabolism of broiler chickens. *Braz J Poult Sci* 2017;19: 83-90.

[66] Al-Harthi MA, Attia YA, El-Shafey AS, Elgandy MF. Impact of phytase on improving the utilisation of pelleted broiler diets containing olive by-products. *Ital J Anim Sci* 2020;19: 310-18.

[67] Al-Harthi MA, Attia YA. Effect of citric acid on the nutritive value of olive cake in broiler diets. *Eur Poult Sci* 2016; 80, 153.

[68] Saleh AA, Paray BA, Dawood MA. Olive cake meal and Bacillus licheniformis impacted the growth performance, muscle fatty acid content, and health status of broiler chickens. *Animals* 2020;10: 695.

[69] Rebollada-Merino A, Ugarte-Ruiz M, Hernández M, Miguela-Villoldo P, Abad D, Cuesta-Álvaro P, et al. Dietary supplementation with fermented defatted "alperujo" induces modifications of the intestinal mucosa and cecal microbiota of broiler chickens. *Poult Sci* 2020;99: 5308-15.

[70] Papadomichelakis G, Pappas AC, Tsiplakou E, Symeon GK, Sotirakoglou K, Mpekelis V, et al. Effects of dietary dried olive pulp inclusion on growth performance and meat quality of broiler chickens. *Livest Sci* 2019;221: 115-22.

[71] Sayehban P, Seidavi A, Dadashbeiki M, Ghorbani A, de Araújo WAG, Durazzo A, et al. Olive pulp and exogenous enzymes feed supplementation effect on the carcass and offal in broilers: a preliminary study. *Agriculture* 2020;10: 359.

[72] Sayehban P, Seidavi A, Dadashbeiki M, Ghorbani A, Araújo WAG, Albino LFT. Effects of different dietary levels of two types of olive pulp and exogenous enzyme supplementation on the gastrointestinal tract size, immunology and hematology of broilers. *Braz J Poult Sci* 2015;17: 73-85.

[73] Branciari R, Galarini R, Giusepponi D, Trabalza-Marinucci M, Forte C, Roila R, et al. Oxidative status and presence of bioactive compounds in meat from chickens fed polyphenols extracted from olive oil industry waste. *Sustainability* 2017;9: 1566.

[74] Nasopoulou C, Lytoudi K, Zabetakis I. Evaluation of olive pomace in the production of novel broilers with enhanced in vitro antithrombotic properties. *Eur J Lipid Sci Technol* 2018;120: 1700290.

[75] Gerasopoulos K, Stagos D, Kokkas S, Petrotos K, Kantas D, Goulas P, et al. Feed supplemented with byproducts from olive oil mill wastewater processing increases antioxidant capacity in broiler chickens. *Food Chem Toxicol* 2015;82: 42-9.

[76] Herrero-Encinas J, Blanch M, Pastor JJ, Mereu A, Ipharraguerre IR, Menoyo D. Effects of a bioactive olive pomace extract from Olea europaea on growth performance, gut function, and intestinal microbiota in broiler chickens. *Poult Sci* 2020;99: 2-10.

[77] Herrero-Encinas J, Blanch M, Pastor JJ, Menoyo D. Diet supplementation with a bioactive pomace extract from *Olea europaea* partially mitigates negative effects on gut health arising from a short-term fasting period in broiler chickens. *Animals* 2020;10: 349.

[78] Afsari M, Mohebbifar A, Torki M. Effects of dietary inclusion of olive pulp supplemented with probiotics on productive performance, egg quality and blood parameters of laying hens. *Annu Res Rev Biol* 2013;4: 198-211.

[79] Zangeneh S, Torki M. Effects of b-mannanase supplementing of olive pulp included diet on performance of laying hens, egg quality characteristics, humoral and cellular immune response and blood parameters. *Glob Vet* 2011;7: 391-398.

[80] Zarei M, Ehsani M, Torki M. Productive performance of laying hens fed wheat-based diets included olive pulp with or without a commercial enzyme product. *Afr J Biotechnol* 2011;10: 4303-12.

[81] Al-Harthi MA, Attia YA. Effect of citric acid on the utilization of olive cake diets for laying hens. *Ital J Anim Sci* 2015;14: 3966.

[82] Ozcan C, Cimrin T, Yakar Y, Alasahan S. Effects of olive cake meal on serum constituents and fatty acid levels in breast muscle of Japanese quail. *S Afr J Anim Sci* 2020;50: 874-80.

[83] Abd El-Moneim AEME, Sabic EM, Abu-Taleb AM. Influence of dietary supplementation of irradiated or non-irradiated olive pulp on biochemical profile, antioxidant status and immune response of Japanese quails. *Biol Rhythm Res* 2019;in press.

[84] Abd El-Moneim AE, Sabic EM. Beneficial effect of feeding olive pulp and *Aspergillus awamori* on productive performance, egg quality, serum/yolk cholesterol and oxidative status in laying Japanese quails. *J Anim Feed Sci* 2019;28: 52-61.

[85] Elfalleh W, Ying M, Nasri N, Sheng-Hua H, Guasmi F, Ferchichi, A. Fatty acids from Tunisian and Chinese pomegranate (*Punica granatum* L.) seeds. *Int J Food Sci Nutr* 2011;62: 200-6.

[86] Dkhil MA. Anti-coccidial, anthelmintic and antioxidant activities of pomegranate (*Punica granatum*) peel extract. *Parasitol Res* 2013;112: 2639–46.

[87] Abdel Baset S, Ashour EA, Abd El-Hack ME, El-Mekkawy MM. Effect of different levels of pomegranate peel powder and probiotic supplementation on growth, carcass traits, blood serum metabolites, antioxidant status and meat quality of broilers. *Anim Biotechnol* 2020;in press.

[88] Ghosh S, Chatterjee PN, Maity A, Mukherjee J, Batabyal S, Chatterjee JK. Effect of supplementing pomegranate peel infusion on body growth, feed efficiency, biochemical metabolites and antioxidant status of broiler chicken. *Trop Anim Health Prod* 2020;52: 3899-905.

[89] Rajani J, Torshizi MK, Rahimi S. Control of ascites mortality and improved performance and meat shelf-life in broilers using feed adjuncts with presumed antioxidant activity. *Anim Feed Sci Technol* 2011;170: 239-45.

[90] Ahmed ST, Islam MM, Bostami AR, Mun HS, Kim YJ, Yang CJ. Meat composition, fatty acid profile and oxidative stability of meat

from broilers supplemented with pomegranate (*Punica granatum* L.) by-products. *Food Chem* 2015;188: 481-8.

[91] Akuru EA, Mpendulo CT, Oyeagu CE, Nantapo CWT. Pomegranate (*Punica granatum* L.) peel powder meal supplementation in broilers: effect on growth performance, digestibility, carcase and organ weights, serum and some meat antioxidant enzyme biomarkers. *Ital J Anim Sci* 2021;20: 119-31.

[92] Bostami ABMR, Ahmed ST, Islam MM, Mun HS, Ko S, Kim S, et al. Growth performance, fecal noxious gas emission and economic efficacy in broilers fed fermented pomegranate byproducts as residue of fruit industry. *Int J Adv Res* 2015;3: 102-14.

[93] Ahmed ST, Ko SY, Yang CJ. Improving the nutritional quality and shelf life of broiler meat by feeding diets supplemented with fermented pomegranate (*Punica granatum* L.) by-products. *Br Poult Sci* 2017;58: 694-703.

[94] Gungor E, Altop A, Erener G, Coskun I. Effect of raw and fermented pomegranate pomace on performance, antioxidant activity, intestinal microbiota and morphology in broiler chickens. *Arch Anim Nutr* 2012;75: 137-52.

[95] Hamady GA, Abdel-Moneim MA, El-Chaghaby GA, Abd-El-Ghany ZM, Hassanin MS. Effect of Pomegranate peel extract as natural growth promoter on the productive performance and intestinal microbiota of broiler chickens. *Afr J Agric Sci Technol* 2015;3: 514-9.

[96] Kishawy AT, Amer SA, Abd El-Hack ME, Saadeldin IM, Swelum AA. The impact of dietary linseed oil and pomegranate peel extract on broiler growth, carcass traits, serum lipid profile, and meat fatty acid, phenol, and flavonoid contents. *Asian-Australas J Anim Sci* 2019;32: 1161-71.

[97] Saleh H, Golian A, Kermanshahi H, Mirakzehi MT. Effects of dietary α-tocopherol acetate, pomegranate peel, and pomegranate peel extract on phenolic content, fatty acid composition, and meat quality of broiler chickens. *J Appl Anim Res* 2017;45: 629-36.

[98] Saleh H, Golian A, Kermanshahi H, Mirakzehi MT. Antioxidant status and thigh meat quality of broiler chickens fed diet supplemented with α-tocopherolacetate, pomegranate pomace and pomegranate pomace extract. *Ital J Anim Sci* 2018;17: 386-95.

[99] Perricone V, Comi M, Giromini C, Rebucci R, Agazzi A, Savoini G, et al. Green tea and pomegranate extract administered during critical moments of the production cycle improves blood antiradical activity and alters cecal microbial ecology of broiler chickens. *Animals* 2020;10: 785.

[100] Sharifian M, Hosseini-Vashan SJ, Nasri MF, Perai AH (2019). Pomegranate peel extract for broiler chickens under heat stress: Its influence on growth performance, carcass traits, blood metabolites, immunity, jejunal morphology, and meat quality. *Livest Sci* 2019;227: 22-8.

[101] Hosseini-Vashan SJ, Raei-Moghadam MS. Antioxidant and immune system status, plasma lipid, abdominal fat, and growth performance of broilers exposed to heat stress and fed diets supplemented with pomegranate pulp (*Punica granatum* L.). *J Appl Anim Res* 2019;47: 521-31.

[102] Hosseini-Vashan SJ, Sharifian M, Piray AH, Fathi-Nasri MH. Growth performance, carcass and blood traits, immunity, jejunal morphology and meat quality of heat-stressed broiler chickens fed urea-treated pomegranate (*Punica granatum* L.) peel. *Anim Feed Sci Technol* 2020;267: 114553.

[103] Manterys A, Franczyk-Zarow M, Czyzynska-Cichon I, Drahun A, Kus E, Szymczyk B, et al. Haematological parameters, serum lipid profile, liver function and fatty acid profile of broiler chickens fed on diets supplemented with pomegranate seed oil and linseed oil. *Br Poult Sci* 2016;57: 771-9.

[104] Szymczyk B, Szczurek W. Effect of dietary pomegranate seed oil and linseed oil on broiler chicken's performance and meat fatty acid profile. *J Anim Feed Sci* 2016;25: 37-44.

[105] Banaszkiewicz T, Białek A, Tokarz A, Kaszperuk K. Effect of dietary grape and pomegranate seed oil on the post-slaughter value

and physicochemical properties of muscles of broiler chickens. *Acta Sci Pol Technol Aliment* 2018;17: 199-209.

[106] Eid Y, Kirrella AA, Tolba A, El-Deeb M, Sayed S, El-Sawy HB, et al. Dietary Pomegranate By-Product Alleviated the Oxidative Stress Induced by Dexamethasone in Laying Hens in the Pre-Peak Period. *Animals* 2021;11: 1022.

[107] Kostogrys RB, Filipiak-Florkiewicz A, Dereń K, Drahun A, Czyżyńska-Cichoń I, Cieślik E, et al. Effect of dietary pomegranate seed oil on laying hen performance and physicochemical properties of eggs. *Food Chem* 2017;221: 1096-103.

[108] Abbas RJ, Al-Salhie KCK, Al-Hummod SK. The effect of using different levels of pomegranate (*Punica granatum*) peel powder on productive and physiological performance of Japanese quail (*Coturnix coturnix japonica*). *Livest Res Rural Dev* 2017;29: 231.

[109] Şenay S, Islim P, Tugay A. Supplementation of natural antioxidants to reduced crude protein diets for Japanese quails exposed to heat stress. *Braz J Poult Sci* 2019;21: 1-14.

[110] Berto BP, Borba HR, dos Santos HLC, Lima VM, Flausino W, Teixeira-Filho WL, et al. Evaluation of pomegranate (*Punica granatum*) pericarp aqueous extract on *Eimeria* spp. from Japanese quails (Coturnix japonica). *Int J Vet Sci Med* 2014;2: 35-40.

[111] Mohdaly AA, Sarhan MA, Smetanska I, Mahmoud A. Antioxidant properties of various solvent extracts of potato peel, sugar beet pulp and sesame cake. *J Sci Food Agric* 2010;90: 218-26.

[112] Kumari P, Kumar K, Kumar S. Effect of dietary supplement of sugar beet, neem leaf, linseed and coriander on growth performance and carcass trait of Vanaraja chicken. *Vet World* 2014;7: 639-43.

[113] Gonzalez-Alvarado JM, Jimenez-Moreno E, Gonzalez-Sanchez D, Lazaro R, Mateos GG. Effect of inclusion of oat hulls and sugar beet pulp in the diet on productive performance and digestive traits of broilers from 1 to 42 days of age. *Anim Feed Sci Technol* 2010;162: 37-46.

[114] Jimenez-Moreno E, Frikha M, De Coca-Sinova A, Garcia J, Mateos, GG. Oat hulls and sugar beet pulp in diets for broilers 1. Effects on growth performance and nutrient digestibility. *Anim Feed Sci Technol* 2013;182: 33-43.

[115] Jimenez-Moreno E, Frikha M, De Coca-Sinova A, Lazaro RP, Mateos GG. Oat hulls and sugar beet pulp in diets for broilers. 2. Effects on the development of the gastrointestinal tract and on the structure of the jejunal mucosa. *Anim Feed Sci Technol* 2013;182: 44-52.

[116] Abdel-Hafeez HM, Saleh ESE, Tawfeek SS, Youssef IMI, Abdel-Daim ASA. Utilization of potato peels and sugar beet pulp with and without enzyme supplementation in broiler chicken diets: effects on performance, serum biochemical indices and carcass traits. *J Anim Physiol Anim Nutr* 2018;102: 56-66.

[117] Abdel-Daim ASA, Tawfeek SS, El-Nahass ES, Hassan AHA, Youssef IMI (2020). Effect of Feeding Potato Peels and Sugar Beet Pulp With or Without Enzyme on Nutrient Digestibility, Intestinal Morphology, and Meat Quality of Broiler Chickens. *Poult Sci J* 2020;8: 189-99.

[118] Petterson D, Razdan A. Effects of increasing levels of sugar-beet pulp in broiler chicken diets on nutrient digestion and serum lipids. *Br J Nutr* 1993;70: 127-37.

[119] Abbas A, Iqbal Z, Abbas RZ, Khan MK, Khan JA, Mahmood MS, et al. In vivo anticoccidial effects of *Beta vulgaris* (sugar beet) in broiler chickens. *Microb Pathog* 2017;111: 139-44.

[120] Selim S, Hussein E. Production performance, egg quality, blood biochemical constituents, egg yolk lipid profile and lipid peroxidation of laying hens fed sugar beet pulp. *Food Chem* 2020;310: 125864.

[121] Alagawany M, Attia A. Effects of feeding sugar beet pulp and Avizyme supplementation on performance, egg quality, nutrient digestion and nitrogen balance of laying Japanese quail. *Avian Biol Res* 2015;8: 79-88.

[122] Del Valle M, Camara M, Torija M-E. Chemical characterization of tomato pomace. *J Sci Food Agric* 2006;86: 1232-36.

[123] Lu Z, Wang J, Gao R, Ye F, Zhao G. Sustainable valorisation of tomato pomace: A comprehensive review. *Trends Food Sci Technol* 2019;86: 172-87.

[124] Ghazi S, Drakhshan A. The effects of different levels of tomato pomace in broilers chick performance. In: *Proceedings of the 12th European Poultry Conference*, Verona, Italy, pp. 10-14, 2006.

[125] Lira RC, Rabello CB, Mohaupt Marques Ludke MDC, Ferreira PV, Lana GRQ, Valerio Lana SR. (2010) Productive performance of broiler chickens fed tomato waste. *Rev. Bras. Zootec.* 2010;39: 1074-81.

[126] Yitbarek MB. The effect of feeding different levels of dried tomato pomace on the performance of Rhode Island Red (RIR) grower chicks. *Int J Livest Prod* 2013;4: 35-41.

[127] Hosseini-Vashan SJ, Golian A, Yaghobfar A. Growth, immune, antioxidant, and bone responses of heat stress-exposed broilers fed diets supplemented with tomato pomace. *Int J Biometeorol* 2016;60: 1183-92.

[128] Wahyuni HI, Yudiarti T, Widiastuti E, Sugiharto S, Isroli I, Sartono TA. The Use of Tomato Waste Juice as an Antioxidant Source for Broiler Chickens. In: *IOP Conference Series: Earth and Environmental Science*, vol. 518, p. 012006, 2020.

[129] Yannakopoulos AL, Tserveni-Gousi AS, Christaki EV. Effect of locally produced tomato meal on the performance and the egg quality of laying hens. *Anim Feed Sci Technol* 1992;36: 53-7.

[130] Dotas D, Zamanidis S, Balios J. Effect of dried tomato pulp on the performance and egg traits of laying hens. *Br Poult Sci* 1999;40: 695-7.

[131] Jafari M, Pirmohammadi R, Bampidis V. The use of dried tomato pulp in diets of laying hens. *Int J Poult Sci* 2006;5: 618-622.

[132] Nobakht A, Safamehr AR. The effects of inclusion different levels of dried tomato pomace in laying hens diets on performance and

plasma and egg yolk cholesterol contents. *J Anim Vet Adv* 2007;6: 1101-6.
[133] Akdemir F, Orhan C, Sahin N, Sahin K, Hayirli A. Tomato powder in laying hen diets: effects on concentrations of yolk carotenoids and lipid peroxidation. *Br Poult Sci* 2012;53: 675-80.
[134] Panaite TD, Nour V, Vlaicu PA, Ropota M, Corbu AR, Saracila M. Flaxseed and dried tomato waste used together in laying hens diet. *Arch Anim Nutr* 2019;73: 222-38.
[135] Salajegheh MH, Ghazi S, Mahdavi R, Mozafari O. (2012). Effects of different levels of dried tomato pomace on performance, egg quality and serum metabolites of laying hens. *Afr J Biotechnol* 2012;11: 15373-9.
[136] Jouzi H. Vali N, Pourreza J. The effects of tomato pulp powder supplementation on performance and some blood parameters in Japanese quail (*Coturnix japonica*). *J Agric Biol Sci* 2015;10: 103-7.
[137] Nikolakakis I, Banakis D, Florou-Paneri P, Dotas V, Giannenas I, Botsoglou N. Effect of dried tomato pulp on performance and carcass characteristics of growing quails. *Arch Geflügelkd* 2004;68: 34-38.
[138] Botsoglou N, Papageorgiou G, Nikolakakis I, Florou-Paneri P, Giannenas I, Dotas V, et al. Effect of dietary dried tomato pulp on oxidative stability of Japanese quail meat. *J Agric Food Chem* 2004;52: 2982-8.
[139] Sahin N, Orhan C, Tuzcu M, Sahin K, Kucuk O. The effects of tomato powder supplementation on performance and lipid peroxidation in quail. *Poult Sci* 2008;87: 276-83.

Chapter Reviewed by:

Assistant Professor Ilias Giannenas, Director of Laboratory of Nutrition, School of Veterinary Medicine, Aristotle University of Thessaloniki, Thessaloniki, Greece.

Assistant Professor Athanasios Pappas, Laboratory of Nutritional Physiology and Feeding, Department of Animal Science, Agricultural University of Athens, Athens, Greece.

In: Bioactive Compounds …
Editors: G. Yildiz and G. İzli
ISBN: 978-1-68507-361-9

Chapter 11

LYCOPENE: A VALUABLE BIOACTIVE MOLECULE

Raquel Guiné[1,*], Maria João Lima[1], David Castelão[1], Charles Okpala[2], Małgorzata Korzeniowska[2] and Sofia Guiné Florença[3]

[1]CERNAS Research Centre, Polytechnic Institute of Viseu, Viseu, Portugal
[2]Faculty of Biotechnology and Food Science, Wrocław University of Environmental and Life Sciences, Wroclaw, Poland
[3]FCNAUP, University of Porto, Porto, Portugal

ABSTRACT

As a response to external stress, to prevent damage caused by reaction oxygen species (ROS) in the plant cells, they generate

* Corresponding Author's Email: raquelguine@esav.ipv.pt.

antioxidant molecules able to help them fight those adverse conditions. Lycopene is a potent antioxidant found mostly in red fruits and vegetables such as tomato, watermelon or grapefruit. The intake of this bioactive compound has been associated with many health benefits, and therefore the objective of this chapter is to discuss the role of lycopene for an improved health, while also highlighting the most valuable natural sources of this molecule.

The results of the review showed that tomato and tomato products are among the richest sources of lycopene, and that some processing operations can in fact enhance the lycopene content and bioavailability, by favoring the transformation of trans into cis forms of lycopene, better absorbed by the human body. Regarding the health effects, it was found scientific evidence of the beneficial action of lycopene on a variety of diseases, including diabetes, obesity, cardiovascular diseases, vascular health, neurodegenerative diseases and cancer.

Keywords: bioactive compounds, carotenoids, lycopene, health, plant sources

INTRODUCTION

The growing concern with physical well-being and health has been reflected in the way we choose the foods present in our dietary patterns. The modern consumers are more informed, opting more for healthy foods, based on the knowledge that some diseases, such as stroke, cancer or atherosclerosis can somehow be minimized by ingesting some compounds, called bioactive or functional compounds.

Carotenoids are natural dyes present in fruits and vegetables such as, for example carrots, tomatoes, spinach, oranges or peaches, among others. Their chemical structure is composed of conjugated double bonds, which are responsible for their color and some of their biological functions.

Carotenoids, along with vitamins, are the most frequently investigated substances as chemo preventive agents, functioning as antioxidants in biological systems [1, 2].

Lycopene is one of the 600 carotenoid pigments found in nature and one of the 25 found in human plasma and tissues [3]. Lycopene is a natural pigment synthesized exclusively by plants and microorganisms, but not by animals [1, 2, 4]. When the plant ripens, it causes a decrease in the chlorophyll content, leaving lycopene and other carotenoids responsible for the bright colors of most fruits (pineapple, orange, lemon, grapefruit, strawberry, tomato) [5].

Lycopene is one of the most effective antioxidants in the carotenoid family, characterized by its activity against biological species reactive to oxygen, and therefore can prevent oxidative damage to cells and tissues [6]. It is considered the carotenoid that has the greatest oxygen scavenging capacity, because in the molecule 11 of the 13 double bonds are conjugated and form a linear structure (Figure 1) [7]. This extensive system of double bond connections is the key to lycopene's biological activity, namely its susceptibility to oxidative degradation [5].

With few exceptions, lycopene is found in plants in its *trans* form. However, lycopene is susceptible to oxidative cleavage as well as to the trans-cis-isomerization process. When exposed to high temperatures, light, catalysts or active surfaces, it can undergo isomerization reactions, and in that case the less stable *cis* forms predominate. The *cis* isomers (Figure 2) of lycopene have different chemical and physical properties than *trans* isomers, namely greater solubility in oils and hydrocarbon solvents [8]. Note that the bioavailability of *cis* isomers is greater than that of *trans* isomers, and hence, the processing of tomatoes can, under certain conditions, improve the bioavailability in tomato derived products as compared to fresh tomatoes [9].

In recent years, lycopene has received considerable scientific attention. Epidemiological studies showed that the consumption of lycopene is correlated with a reduction in the risk of certain types of cancer as well as cardiovascular disease and mortality in general. *In vivo* and *in vitro* experiments have shown that lycopene in certain amounts (0.5-2 μM) can inhibit the proliferation of cancer cells, induce apoptosis and suppress metastases [9]. Dietary supplementation with lycopene has been associated with a reduced risk of cancer, cardiovascular disease and other diseases (bot in humans and animals) [10].

To this point, the adherence to the Mediterranean diet, which includes frequent consumption of tomatoes, in association with other vegetables, fruits and legumes, and lipids such as olive oil [11], can be a factor that improves general health status and helps prevent certain diseases.

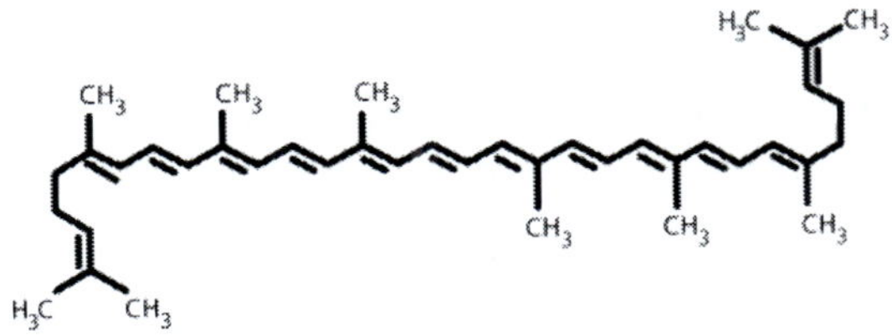

Figure 1. Chemical structure of lycopene, formula: $C_{40}H_{56}$.

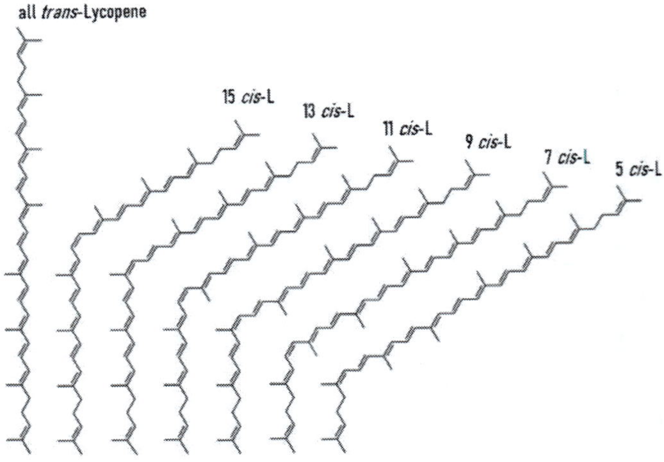

Figure 2. Lycopene forms: *trans* and *cis*.

Dietary Sources of Lycopene

Because the human body is not able to promote the synthesis of carotenoids, the lycopene is obtained through dietary intake. Unlike other carotenoids, lycopene is present in a short list of plant foods that have a red-orange color, such as tomatoes and their derivatives, papaya, red guava and watermelon [12]. Table 1 shows some natural sources of lycopene and Figure 3 shows the contents of lycopene on some processed tomato products.

The amounts of lycopene vary not only according to the type of food, but they may also fluctuate in the same food according to the influence of several factors, such as: the state of ripening, variety, season, growing conditions, characteristics of the soil, climatic conditions and post-harvest changes [12].

Table 1. Natural sources of lycopene [2]

Product	Lycopene (μg/g fresh mass)
Guava	47.0 – 90.0
Watermelon	23.0 – 72.0
Papaya	20.0 – 53.0
Carrot	6.5 – 7.8
Pumpkin	3.8 – 4.6

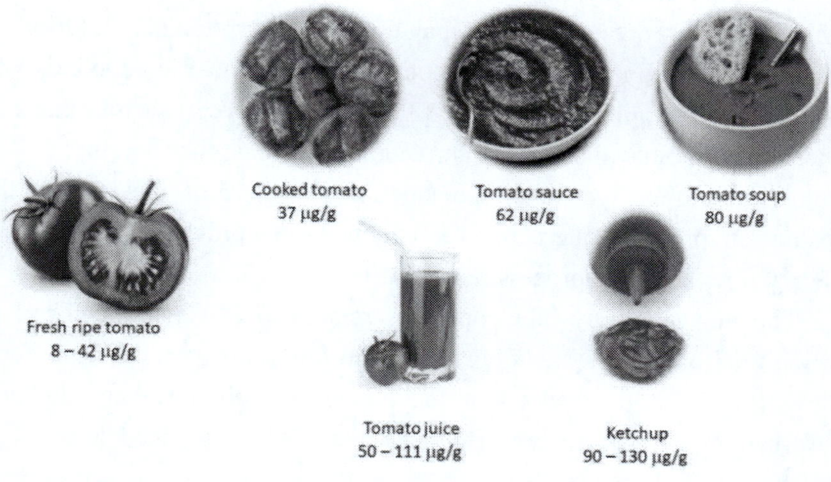

Figure 3. Contents of lycopene in fresh tomato and in some processed tomato products.

It is often found that the intensity of the red color in foods is directly associated with its lycopene concentration. Food peels also present higher levels of lycopene in relation to the pulp, as well as the food produced in regions with a hot climate, when compared to those grown in cooler climates [3, 12, 13]. The proportion of lycopene distributed between all the *trans* and *cis* isomers varies significantly according to the location. Fully *trans* lycopene is predominant in tomatoes and their derived foods and is responsible for approximately 90% of the total lycopene, with the remainder distributed among *cis* isomers. In the industrial processing of tomato products heat treatments are usually employed and therefore in the transformed products the predominance is for the *cis* forms of lycopene. However, during the storage of those products, there is a reversal in isomerization passing again from *cis* isomers, less stable, to their more stable form, that is, *trans* [5].

Bioavailability of Lycopene in the Human Body

The bioavailability of the food constituents is a complex process, which involves digestion, intestinal uptake, absorption, distribution to tissues and the final use by the body tissues of those compounds [2]. According to Moritz & Tramonte [3], there are several factors that can interfere with bioavailability of carotenoids, such as: the food matrix, the isomeric form of lycopene, the amount and type of dietary fat, the absorption process, interactions between carotenoids, the presence of dietary fiber and food processing operations.

The bioavailability of lycopene is related to the isomeric forms in which it is found. Despite being present in food mostly in the *trans* form (80% to 97%), it appears that the isomers in *cis* form are the best absorbed in the human body human, due to their reduced length and better solubility in micelles. Therefore, the isomers in the *cis* form of lycopene have less difficulty in displacement across plasma membranes. Lycopene can be found in human plasma at a 1:1 ratio between *cis* and *trans* isomers. Among the various isomers of lycopene, the isomers *5-cis*,

9-cis, *13-cis* and *15-cis* are those that the human body finds it easier to absorb, for as opposed to the *trans* isomers that occur most frequently in nature [14].

Lycopene ingested in its natural form (*trans*-lycopene), is poorly absorbed. It has been suggested that linear isomers in *trans* form may aggregate within the intestine and form crystals, reducing their absorption by micelles [14]. However, studies show that the thermal processing of tomatoes and their derivatives improves their bioavailability. Thermal processing breaks the cell wall and allows the extraction of lycopene from chromoplast [15].

The acidic pH of the stomach seems to contribute, to some extent, to the transformation of the *trans* form to *cis* lycopene isomers [3].

During digestion and absorption, lycopene is separated from the remaining nutrients, and it is possible that during this separation lycopene isomerization occurs, changing its configuration from *trans* to *cis*-isomers. [16].

The increase in the bioavailability of lycopene in processed products occurs because the heat treatment breaks cell walls, weakening the binding of lycopene in the tissue matrix, facilitating its isomerization to the *cis* form. However, care must be taken with the time of exposure to high temperatures, because in addition to the destruction of the complex protein where lycopene is inserted, lycopene itself can be destroyed. Hence, processing parameters, and particularly the time/temperature binomial, must be optimized [5].

Health Effects of Lycopene

Several researches [17–19] have shown that lycopene can bring several benefits for human health. Although the first results have been promising, subsequent research on the biological effects of lycopene has been sparse. More recently, some works in cell cultures and epidemiological studies in humans and animals sought to investigate more consistently the possible beneficial effects of lycopene. These epidemiological studies quickly revealed the existence of a strong association between increased consumption of tomatoes and derived by-products and reduced risk of some types of disease.

Effect on Cardiovascular Diseases and Cholesterol

Cardiovascular diseases are associated with oxidative stress, inflammatory diseases and vascular dysfunction, and evidence shows the positive effect of lycopene on serum lipid concentrations and blood pressure [20, 21].

When entering the scope of cardiovascular diseases, there are numerous important dietary factors, for this reason a great emphasis has been placed on carotenoids and antioxidant nutrients, such as vitamins E and C. Interest in lycopene continues to grow after the publication of epidemiological and clinical studies that associate high levels of lycopene in the blood with reductions in the incidence of cardiovascular disease [22, 23].

Gajendragadkar et al. [24] evaluated the effects of lycopene in patients with cardiovascular disease by oral administration of 7 mg of lycopene to 36 individuals with cardiovascular disease, using statins, and 36 healthy participants, for 2 months. The study showed that vasodilation improved by 53% in the sick group who received lycopene supplementation. Therefore, the administration of lycopene improved the endothelial function in patients with cardiovascular diseases.

The EURAMIC[1] study, which assessed myocardial infarction in men among 10 European countries, found that the concentration of lycopene in adipose tissue is independently protective against myocardial infarction. The protective effect of lycopene was stronger among non-smokers and in subjects who had higher levels of polyunsaturated fat in adipose tissue [25].

---

[1]EURAMIC is the European community multicenter study on antioxidants, myocardial infarction and breast cancer.

Other studies concluded that lower plasma lycopene levels were associated with early atherosclerosis, manifested as an increase in the thickness of the intima-media wall of the carotid artery [28]. The Kuopio Ischaemic Heart Disease Risk Factor Study showed an average reduction of 4% in the risk of acute coronary events for each 10 nmol/L increase in serum lycopene [28].

Rissanen et al. [28] studied the relationship between the serum concentration of lycopene and the average carotid artery thickness in 1028 men between 46-64 years old, who were classified in 4 groups, according to the lycopene concentrations. The results showed that the low concentration of serum lycopene was associated with an increase in the thickness of the carotid artery, suggesting that lycopene can play an important role in the early stages of atherosclerosis.

Lycopene's anti-atherogenic effect was evaluated in 40 adult male rabbits, with a diet rich in fat, administering 4 mg/kg (during 4 weeks) and 12 mg/kg (over 8 weeks) of lycopene, respectively [29]. Morphological analysis revealed that lycopene reduced the formation of atherosclerotic plaques in the aorta compared with the situation of rabbits with a high-fat diet without lycopene supplementation.

Ried and Fakler [30] evaluated the protective effect of lycopene on serum cholesterol and blood pressure, by means of a meta-analysis of intervention trials, searched on PubMed and Cochrane databases. It was concluded that lycopene administered in doses of 25 mg per day or over was effective in reducing low density lipoproteins (LDL) cholesterol by approximately 10%.

McEneny et al. [31] investigated the ability of lycopene to decrease the high density lipoprotein (HDL) cholesterol in middle-aged individuals with overweight, through the administration of lycopene for a period of 12 weeks. The results showed that the intake of lycopene, in middle-aged individuals, protects the heart through the change in HDL lipoprotein, and therefore the authors claim that lycopene has antiatherogenic properties.

Effect on Male Infertility

Lycopene affects various biological processes of the prostate, such as signaling, cellular communication and progression [32]. So far, few studies have examined the effects of lycopene supplementation on infertility [19].

In a clinical trial, the effects of lycopene supplementation on the parameters and function of sperm were investigated using two groups of men with oligozoospermia. The first group received 25 mg of lycopene supplement for 12 weeks, the second group received doses of placebo for the same period of time. In the lycopene group, there was a significant improvement in volume, concentration, total count, total mobility and levels of total antioxidant capacity, while in the placebo group, no significant changes were observed except for total mobility. However, no significant changes were observed in the mobility and morphology of sperm among them [33].

In addition, it is considered that structural and functional damage to the testis, as well as changes in the quality of sperm due to oxidative stress, can be prevented with lycopene supplementation [34].

Effect on Diabetes and Obesity

Several clinical studies have found an inverse association between lycopene levels and body mass index (BMI). In a study conducted by Park et al. [35], the results revealed that BMI was negatively correlated with serum levels of lycopene in overweight children aged 9 to 10 years. These findings were further corroborated by a study of 374 cases carried out by Sluijs et al. [36], in which the consumption of lycopene resulted in a decline in visceral and subcutaneous fat, as well as in waist circumference in men aged 40-80 years.

In a control study it was found that a combination of anti-inflammatory nutritional supplements (including lycopene), maintained high molecular weight adiponectin concentrations and reduced HOMA-IR (Homeostatic Model Assessment for Insulin Resistance) [37].

In another study made with middle-aged adults with moderate obesity, lycopene supplementation increased the relative abundance of beneficial intestinal microbacteria, bifidobacteria and lactobacilli and improved hepatic lipid metabolism and skeletal muscle oxygenation, as well as increased sebum viscosity cutaneous and reduced exfoliation of skin corneocytes [38].

In addition, tomato juice has been shown to reduce body weight, body fat, waist circumference and BMI in young women (20-30 years) with BMI $\geq$ 20 [39]. Likewise, tomato juice provided antioxidant and anti-inflammatory benefits, as evidenced by increased plasma total antioxidant capacity, erythrocyte superoxide dismutase (SOD), catalase (CAT) and glutathione peroxidase (GSH-Px), and reduced serum levels of MCP-1 (monocyte chemoattractant protein-1), malondialdehyde, thiobarbituric acid reactive substances, interleukin-8 and tumor necrosis factor-α [40].

More recently, a clinical trial has shown that tomato juice enriched with lycopene improved glucose and lipid metabolism by attenuating oxidative stress through the regulation of the mitochondrial metabolism of T lymphocytes in obese children [41].

The accumulation of evidence from epidemiological studies supports the notion that the consumption of lycopene in the diet can help reduce the risk of obesity and diabetes. Therefore, prospective clinical studies are needed to clarify the relationship between lycopene intake and the development of obesity and diabetes. For this purpose, researchers should consider specific populations, accurate measurement of lycopene consumption, clear definition of obesity and diabetes, updated physical exams and other confounding factors [42].

Effect on Depression

Advanced oxidative stress or deficit in antioxidant defenses are related to the pathogenesis of depressive symptoms [43–45].

The relationship between different vegetables was investigated, including tomatoes/tomato products (one of the main sources of lycopene) and the symptoms depressive people based on an elderly population community [46]. The results revealed that incidence of mild depressive symptoms was about 35% and severe was 20%. The authors observed that there was an inverse relation between having mild and severe depressive symptoms when increasing the quantities of tomatoes/tomato products in the diet. Contrarily, no relation was observed between the incidence of depressive symptoms and the intake of other kinds of vegetables. This study confirmed that a tomato-rich diet is associated with lower prevalence of depressive symptoms, and therefore could be used as tool to prevent depressive symptoms, although further studies are still necessary to confirm these assumptions.

Effect on Neurodegenerative Diseases

Tomato is an important source of lycopene and there is a strong association between its intake and blood concentrations of lycopene [47]. Culinary practices promote the isomerization of lycopene and increase its bioavailability. Due to the lipophilic nature of lycopene, consuming tomatoes in combination with lipids further increases lycopene absorption [48]. So far, promising results on the neuroprotective action of tomato/lycopene have been largely confirmed in pre-human research. Lycopene protects lipoproteins and vascular cells from oxidation, being an effective inhibitor of singlet oxygen, having its effect on plasma, low-density lipoproteins and human lymphoid cells *in vitro* [49]. Based on some reports, lycopene's effectiveness as well as its antioxidant activity are ten times greater when compared to vitamin E [49]. Lycopene has shown significant inhibitory effects on neurotoxicity induced by neurotoxins and other toxic factors. Lycopene has also been found to protect the hippocampal neurons of rats treated with trimethyl tin (TMT) (a compound with neurotoxic effects used to model neurodegenerative diseases), reducing the accumulation of reactive oxygen species in the mitochondria during the time of exposure and apoptosis [50]. *In vitro* studies have shown neuroprotective effects of lycopene also on neurotoxicity induced by other substances, including monosodium glutamate (MSG) and t-BHP (tert-Butyl hydroperoxide), or pollutants such as methylmercury (MeHg) and cadmium (Cd).

The concentrations of lycopene in the tissues of the human body, particularly in the liver, lungs, prostate, colon and skin are higher than those of other carotenoids [51]. Lycopene can cross the blood-brain barrier and is detectable in the human brain, suggesting a possible role for lycopene consumption in maintaining neuronal health or in preventing or mitigating neurological injuries induced by oxidative stress [52].

To date, research in the field of neuro-nutrition can open interesting windows on effective and safe treatments, even in patients with dementia or individuals with multimorbidity. Studies indicate that tomatoes are a potentially useful product without any serious side effects and, with promising use in the prevention and treatment of Alzheimer's disease as well as other dementias. The administration of tomato products in humans may be useful to increase serum levels of lycopene, reduce oxidative stress, attenuate/prevent neuro-degeneration processes and delay the onset of dementia. The prophylactic and/or therapeutic value of lycopene or substances containing lycopene in Parkinson's disease is also a promising field to be further evaluated in scientific studies.

Alzheimer's Disease

Alzheimer's disease is the most expressive cause of dementia and is rapidly assuming the position of one of the most expensive, lethal, and burdening diseases of the 21st century [53].

The earliest phase of Alzheimer's disease, corresponding to the cellular phase, happens in parallel with the accumulation of amyloid β, inducing the spread of tau pathology.

In vitro studies of Alzheimer's disease showed that pretreatment with lycopene reduced cell damage induced by exposure to Aβ 1-42 (β-amyloid 1-42; peptides with 36-43 amino acids being the main constituent of the amyloid plaques observed in the brain of patients with Alzheimer's disease). It also prevented cell apoptosis stimulated by Aβ 1-42, through inhibition of the production of reactive oxygen species and reduction in mitochondrial dysfunction [50], by decreasing the expression of apoptotic pro-factors [54] and suppressing Aβ 1–42 activation of NF-κB (nuclear factor kappa-light-chain-enhancer of activated B cells; protein complex that controls DNA transcription, production of cytokines and cell survival), thereby inhibiting the pro-inflammatory response in the microglia [55].

In animal models with Alzheimer's disease, chronic lycopene administration delayed Aβ 1–42-induced paralysis in *Caenorhabditis elegans* [56] and was able to reduce the effects of intracerebral injection of Aβ 1 –42 in rat brains, including mitochondrial dysfunction, a wave of pro-inflammatory cytokines TNF-α, TGF-β (Transforming growth factor beta) and IL-1β, as well as NF- κB and caspase-3 activity [57]. Lycopene supplementation was observed with the purpose of reversing the accumulation of Aβ 1–42 induced by lipopolysaccharides (LPS) and the increase in BACE1 (beta-site amyloid precursor protein cleaving enzyme 1), with positive effects corresponding to an improvement in the performance of memory [58].

In transgenic "tau" rats (rats that express the P301L mutation), dietary supplementation of lycopene improves cognitive performance, inhibiting oxidative stress, increasing the activity of glutathione peroxidase (GSH-Px) and decreasing serum levels of malondialdehyde (MDA). A more significant action was observed after the administration of a treatment composed of lycopene + vitamin E [59]. In humans with Alzheimer's disease, plasma lycopene concentrations are significantly lower than in healthy individuals of the same age [60] and inversely correlated with levels of 8-hydroxy-20-deoxyguanosine (8-OHdG; a marker nucleic acid oxidation), which increase progressively with aging and are higher in patients with Alzheimer's disease [60].

The neuroprotective effect of lycopene was investigated against mitochondrial oxidative stress, in induced Alzheimer's disease. Lycopene in doses of 2.5 and 5 mg/k was administered to male Wistar rats for 21 days. [61]. The administration of lycopene improved the retention of memory, reduced oxidative damage and neuroinflammation, suggesting the effect of lycopene on patients with Alzheimer's disease.

Parkinson's Disease

Parkinson's disease is a devastating neurodegenerative disease that affects millions of people worldwide. A pathological alteration characterized in Parkinson's disease is the degeneration of dopaminergic neurons, which causes a decrease in dopamine in the striatum, as well as the development of motor deficiency [62].

A daily dose of lycopene (20 mg/kg) for 7 days has been shown to reduce the increase induced by MPTP (1-methyl-4-phenyl-1,2,3,6-tetrahydropyridine) in various apoptotic markers in the rat striatum, including Bax and caspase, suggesting that the effect of lycopene in the fight against Parkinson's disease can be mediated by the reduction of neuronal apoptosis [63].

The administration of lycopene-enriched tomato powder can prevent MPTP-induced decrease in dopamine in a Parkinson's model in rats and the implementation of the tomato diet can suppress 6-OHDA (6-hydroxydopamine) induced dopamine decrease [64]. However, it remains to be seen how lycopene inhibits neuronal apoptosis. Further research is needed to elucidate whether the antioxidant and anti-apoptotic effect of lycopene in models of Parkinson's disease is interconnected and whether the regulation of the neuroinflammatory response may contribute to the therapeutic effect of lycopene.

Cerebral Ischemia

Cerebral ischemia is a major cause of cerebrovascular diseases [65].

Lycopene has a protective effect on focal cerebral ischemia in rats since it mediates at least part of the free radical scavenging activity and inhibits microglia activation, reducing the expression of inflammatory cytokines such as TNF-α (Tumor necrosis factor), IL-1β (Interleukin 1 beta) and ICAM-1 (Intercellular Adhesion Molecule 1) [66], with subsequent reduction in the volume of brain damage due to ischemia/reperfusion [67]. The administration of lycopene after subarachnoid hemorrhage can limit neurological deficits resulting from the injury, improve cerebral edema, permeability of the blood-brain barrier and reduce neuronal apoptosis [66].

Oral administration of lycopene (5 mg in 100 g of food, for 1½ to 2 months) has been shown to increase the level of Bcl-2 (B-cell lymphoma 2) and to reduce the activity of caspase-3, a cysteine- protease involved in ischemia-induced neuronal death, causing DNA fragmentation, in ischemic brain tissues in Mongolian squirrels [68], suggesting that lycopene may protect neurons against ischemic damage by inhibiting neuronal apoptosis. In addition, the administration of lycopene can reverse the decrease in levels of superoxide dismutase (SOD) in the hippocampus induced by ischemia in Mongolian squirrels [68]. Since SOD acts mainly to eliminate the harmful effect of oxidative damage, studies indicate that improving the antioxidant capacity of the ischemic neuron may increase the protective effect of lycopene in cerebral ischemia. This hypothesis is supported by the results of studies in which pretreatment with lycopene at a dose of 4 mg/kg reduces the volume of cerebral infarction by eliminating free radicals and inhibiting iron-catalyzed lipid peroxidation in a model of rats with cerebral ischemia induced by occlusion of the middle cerebral artery (MCAO) [67].

Subarachnoid Hemorrhage

Subarachnoid hemorrhage (SAH) is a devastating disease of the Central Nervous System (CNS). Its main detrimental effects on brain function include rupture of the blood-brain barrier (BBB) and cerebral edema [69]. These pathological changes must be treated efficiently in order to minimize the neuronal damage induced by SAH.

A study reported that acute treatment with lycopene (2 h, 40 mg/kg) after building the SAH model improved SAH-induced interruption in BBB, increased brain water content and neurological deficits in SD rats

(Sprague Dawley) through attenuation of neuronal apoptosis and proinflammatory response in the cortex [66], suggesting that lycopene may have potential values for the treatment of SAH.

Epilepsy

Epilepsy is one of the most prevalent neurological diseases in the CNS. It affects approximately 0.5-1.0% of the population, without effective prevention and/or therapeutic strategies to date in about one third of patients [70]. Oxidative stress mediated by mitochondria is an important mechanism for the onset of crises [71]. It can promote the production of ROS and the oxidation of proteins/lipids/nucleotides, thus increasing the susceptibility of the brain to pro-epileptic stimuli through the compromise of mitochondrial bioenergetics [71]. Targeting oxidative stress mediated by mitochondria can be beneficial for the treatment of epilepsy.

A study carried out demonstrated that long-term treatment with lycopene (29 days) in a dosage of 5 and 10 mg/kg promoted the prevention of epilepsy in rats by reversing oxidative damage induced by PTZ (pentylenetetrazole), as well as impairment of the activities of the mitochondrial enzyme complex I, II and IV [17]. In addition, 14 days of treatment with lycopene (2 mg/kg) were found to attenuate PTZ-induced seizure activity by reversing the PTZ-induced increase in brain thiobarbituric acid reactive substance levels, as well as PTZ-induced decrease in the reduction of GSH, SOD, CAT and levels of gamma-aminobutyric acid (GABA) in Swiss albino rats [72].

Although these effects have been observed in different types of rats, they strongly indicate that lycopene can be recommended for the medical treatment of epilepsy, because lycopene has powerful antioxidant properties and is widely used as a dietary supplement. However, the real effects of lycopene in patients with epilepsy should be evaluated in future studies [73].

Ethanol-Induced Congnitive Disfunction

Excessive exposure to alcohol is a common problem in modern society. It can induce reduced brain volume and cognitive dysfunction [74]. Numerous mechanisms, such as promoting microglia-mediated neuroinflammation and increasing the sensitivity of the NMDA receptor (N-methyl-D-aspartate), have been discovered to explain the reason for alcohol-induced neurotoxicity [75]. Nevertheless, the authors still do not know exactly how the structure and function of the brain is affected by ethanol [75]. Recent studies on the regulation of human astrocytes by lycopene may provide potential answers to this question [76]. In these studies, pretreatment with lycopene (5 and 10 μM) was used to reverse the ethanol-induced increase in poly (ADP-ribose) levels, as well as ethanol-induced decrease in nicotinamide and adenine (NAD^+) dinucleotide levels in human U251 astroglioma cells and primary human astrocytes [76]. The accumulated poly(ADP-ribose) reflects a high activity of the poly(ADP-ribose) polymerase (PARP), which also contributes to the reduction in the concentration of ATP due to the depletion of NAD^+ [77]. The effect of lycopene on restoring NAD^+ levels in human astrocytes not only causes astrocytes to acquire complete ATP, but helps maintain sirtuin 1 (SIRT1) activity [76]. Since the regulation of SIRT1 by NAD^+ has been reported to impact the course of neurodegenerative diseases, the regulation of SIRT1 activity by lycopene in human astrocytes may have important values for understanding the mechanism of action of lycopene in ethanol-induced brain dysfunction [78].

Anti-Cancer Properties

Cancer is a group of progressive diseases that goes through several stages of accumulation of mutant genes, cell transformation, morphological changes and high rates of proliferation. Malignant cancer cells are characterized by persistent proliferation, evasion of apoptosis, escape from immune destruction, migration and invasion, angiogenesis and energy deregulation [79].

Lycopene is an efficient inhibitor of cell proliferation, and the different effects observed under various conditions could be determined by the concentration of lycopene present in the site. Lycopene is well distributed in many tissues of the body, the liver being the organ where it accumulates the most [80].

The high intake of lycopene ($\geq$ 3.42 mg/day) or tomato (> 1.95 g/day) can significantly decrease the mortality rate from stomach, colon and liver cancer [81]. Other cancers whose risk seems to be inversely associated with serum levels or lycopene tissue cells include that of the breast, digestive system, cervix, bladder, skin and possibly lung.

The proposed mechanisms through which lycopene can influence the risk of cancer are related to its functioning as antioxidant. The antioxidant function of lycopene may also explain the observation in a multicentre European study in which carotenoid levels in adipose tissue were inversely associated with the risk of myocardial infarction [82, 83].

It was found that the synergistic interaction of lycopene with other nutrients in the diet increases their action in the anti-cancer or chemoprevention process, thus validating the vital role of these dietary interventions in cancer prevention [84]. According to lycopene's antioxidant property, several researchers have combined it with some anti-cancer drugs and phyto-nutrients from various products in order to improve its effectiveness [85]. Recent research has shown that lycopene can increase the effect of anti-cancer drugs (including adriamycin, cisplatin, docetaxel and paclitaxel) on the growth of cancer cells and also play a role in reducing tumor size through the mediation of oncogenetic signaling pathways and regulation of the redox potential.

Breast Cancer

Some epidemiological and clinical trials have shown a negative correlation between blood lycopene levels and the risk of breast cancer [86].

Lycopene assumes an anti-proliferative mechanism by selectively inhibiting the growth of cancer cells, not acting on normal cells. *In vitro* results obtained demonstrated that lycopene has no adverse effects on normal monolayer or organotypic oral mucosa cells and selectively restricts cell proliferation in breast cancer cells in humans [87]. Lycopene also improved the regulation of cell factor signaling in breast cancer cells by topoisomerase inhibitors without damaging normal breast cells [88]. The specific mechanism of lycopene's anti-proliferative activity and its metabolites in cancer cells occurs mainly through the regulation of signaling that induces cell cycle termination, affecting intercellular communication and inhibiting cell survival [89].

In breast and endometrial cancer cell lines, lycopene treatment reduced serum-induced phosphorylation of Rb proteins (retinoblastoma), decreased levels of D1 and D3 cyclin and inhibited cell cycle progression [90]. Another *in vitro* trial demonstrated that lycopene inhibited ultraviolet-B-induced superproliferation of human keratinocytes and reduced the CDK2 and CDK4 (Cyclin Dependent Kinase), cyclin E and cyclin D complexes [18].

Prostate Cancer

Epidemiological and clinical trials have demonstrated an inverse association between lycopene and the risk of prostate cancer [91].

Benign prostatic hyperplasia (BPH) is a common prostate disease that affects men as they age and a risk factor for the development of prostate cancer. Some studies have shown that lycopene has an antimetastatic effect and can be considered an antiproliferative and anticancer agent, inhibiting cell adhesion, invasion and migration, thus reducing the occurrence or the progression of prostate cancer [92].

The beneficial role of lycopene in patients with BPH was studied, by conducting a pilot study in 40 elderly men and also in a placebo group, who were randomized to receive 15 mg/day of lycopene, for 6 months. The results showed that lycopene supplementation decreased levels of PSA (prostate specific antigen), and there were no changes in the placebo group. Hence it was concluded that lycopene can play an inhibiting role in progression of BPH [93].

Changes in normal epithelial prostate cells were investigated, for the administration of synthetic lycopene (up to 5 mol/L), and the results indicated that the administration of lycopene inhibited the growth of normal prostate cells, *in vitro*. This fact may have an effect on the prostate development and/or on the enlargement of the prostate tissue, such as that found in patients with benign prostatic hyperplasia (BPH), a potential precursor to prostate cancer [94].

The results of the largest prospective studies suggest that dietary lycopene is protective against prostate cancer. However, molecular and genetic differences can result in varied responses to carotenoids [95].

Elgass et al. [96] consulted an *in vitro* study of the effect of lycopene on cell adhesion and the migration properties of cancer cells in prostate cancer. They concluded that lycopene inhibited the adhesion of cancer cells and influenced the migratory properties of carcinogenic cells.

The anti-tumor effect of lycopene was evaluated alone or in combination with chemotherapeutic agents in the prostate cancer. The study concluded that lycopene influences the cell proliferation, increases the anti- proliferative and chemotherapeutic effects, increases the anti-proliferation of agonists, modulates the progression of the cell cycle, enhances chemotherapeutic agent-induced apoptosis, and alters the expression of Biomarker genes in prostate cancer [97].

Oral Cancer

Oral malignancy occupies the twelfth position among all cancers and has a mortality of 50%, with no significant improvement over the past 30 years. The potential importance of lycopene in the development, prevention and treatment of pre-malignant lesions and oral cancer was investigated by Lu et al. [98] through preliminary clinical trials in humans. The results demonstrated that lycopene has a therapeutic effect on precancerous lesions, being a promising anticancer agent against mouth cancer.

Livny et al. [99] studied the interference of lycopene in the processes of cancer cells derived from the oral cavity. For this, cells from an oral cavity tumor were incubated with different concentrations of lycopene, through the cell culture medium. It was observed an inhibition in the proliferation of oral tumor cells, being this dependent on the dose of lycopene. Therefore, it was concluded that lycopene can be an effective anti-cancer agent in oral cancer.

Gastric Cancer

Gastric cancer is the second most common cause of cancer death in the world. Epidemiological studies have examined the possible relationship between tomato products and gastric cancer and a meta-analysis was made by Yang et al. [100] of cohort and case-control studies to analyze this association. They concluded that the consumption of high quantities of tomato products was in fact associated with a decrease in the risk of developing gastric cancer. Nevertheless, due to the possible influence of confounding factors and exposure misclassification, they recommend that additional studies should be implemented to establish a definite association between the ingestion of lycopene rich foods and gastric cancer.

Renal Cancer

Renal cell carcinoma (RCC) is the most frequent upper urinary tract cancer in humans and accounts for 80-85% of malignant renal tumors [101].

The consumption of lycopene has been associated with a decreased risk of development of cancer in kidney cells. This activity may be associated due to its potent antioxidant activity and anti-inflammatory effects.

One study by Cross et al. [102] also investigated the role of a diet rich in lycopene in the development of cancer in renal cells in the Eker mouse model with mutation TSC2. The results highlighted that dietary supplementation with lycopene reduces the development of renal cell cancers in the predisposed TSC2 mutant Eker rat model. Hence, the authors postulate that lycopene may have a prominent role in the prevention of RCC. A similar study conducted by the same team, with the same rat model (Eker TSC2 mutant) [101] concluded that the incidence of renal carcinomas was statistically significantly decreased in the lycopene-treated rats in comparison with the untreated controls. Additionally, tumor length decreased also in lycopene supplemented rats.

CONCLUSION

Lycopene has been associated with a potent antioxidant capacity, which brings many benefits for the human body, especially those who are associated with oxidation by reactive oxygen species. Among these, stand specially the neurodegenerative diseases and cancer. Many studies have established an inverse correlation between ingestion of lycopene and the development of this type of disease.

So far, the best recommendation in terms of behavior and nutrition is the adoption of an active lifestyle, combined with eating habits according to the Mediterranean diet, which includes the frequent consumption of ripe tomatoes, highly rich in lycopene, in association with lipids such as olive oil.

Nevertheless, it is still necessary to proceed with well-designed research, with a clinical focus, that may add to the existing scientific evidence the confirmation that the regular consumption of lycopene may in fact be a fundamental component of a diet aimed at promoting cardiovascular and brain health, fighting cancer and other health problems.

ACKNOWLEDGMENTS

The authors would like to acknowledge FCT - Foundation for Science and Technology, I.P., within the scope of the project Ref[a]

UIDB/00681/2020. Furthermore, we would like to thank the CERNAS Research Centre and the Polytechnic Institute of Viseu for their support.

The authors thank the valuable advice of the experts who revised the present chapter: Professor Aida Moreira da Silva (Lic, MSc, PhD), R&D Unit in Molecular Chemistry-Physics, Department of Chemistry, University of Coimbra, Portugal and Polytechnic Institute of Coimbra, Coimbra College of Agriculture, Portugal; Professor Monica Tarcea (MD, PhD, Habil), Department of Community Nutrition and Food Safety, University of Medicine, Pharmacy, Science and Technology, Targu Mures City, Romania.

References

[1] Machado, CX. *Tomate – o papel do licopeno na proteção antioxidante* [*Tomato - the role of lycopene in antioxidant protection*]. Brasil: Universidade Federal do Rio Grande do Sul., 2005.

[2] Marques, CS; dos, S. O licopeno como composto bioativo do tomate. Dissertação de Mestrado. [Lycopene as a tomato bioactive compound. Masters dissertation.] Viseu, Portugal: Instituto Politécnujco de Viseu, 2015.

[3] Moritz, B; Tramonte, VLGC. Biodisponibilidade do licopeno [Lycopene bioavailability]. Revista De Nutricao-Brazilian *Journal of Nutrition*, 2006, 19, 265–73.

[4] Miranda, KF. *Estudo da concentração de licopeno por ultrafiltração a partir de suco de melancia* (*Citrullus vulgaris Schard*) [*Study of lycopene concentration by ultrafiltration from watermelon juice* (*Citrullus vulgaris Schard*)]. Brasil: Unicamp, 2005.

[5] Shi, J; Le Maguer, M. Lycopene in tomatoes: chemical and physical properties affected by food processing. *Crit Rev Biotechnol*, 2000, 20, 293–334.

[6] Boeira, SP; Filho, CB; Del'Fabbro, L; Roman, SS; Royes, LFF; Fighera, MR; et al. Lycopene treatment prevents hematological, reproductive and histopathological damage induced by acute zearalenone administration in male Swiss mice. *Exp Toxicol Pathol*, 2014, 66, 179–85.

[7] Anguelova, T; Warthesen, J. Degradation of Lycopene, α-Carotene, and β-Carotene During Lipid Peroxidation. *J Food Sci*, 2000, 65, 71–5.

[8] Rao, AV; Agarwal, S. Role of antioxidant lycopene in cancer and heart disease. *J Am Coll Nutr*, 2000, 19, 563–9.

[9] Cheng, HM; Koutsidis, G; Lodge, JK; Ashor, A; Siervo, M; Lara, J. Tomato and lycopene supplementation and cardiovascular risk factors: A systematic review and meta-analysis. *Atherosclerosis*, 2017, 257, 100–8.

[10] Mordente, A; Guantario, B; Meucci, E; Silvestrini, A; Lombardi, E; Martorana, GE; et al. Lycopene and cardiovascular diseases: an update. *Curr Med Chem*, 2011, 18, 1146–63.

[11] De, Amicis, R; Leone, A; Foppiani, A; Osio, D; Lewandowski, L; Giustizieri, V; et al. Mediterranean Diet and Cognitive Status in Free-Living Elderly: A Cross-Sectional Study in Northern Italy. *J Am Coll Nutr*, 2018, 37, 494–500.

[12] Shami, NJIE; Moreira, EAM. Licopeno como agente antioxidante [Lycopene as an Antioxidant Agent]. *Revista de Nutricao* 2004, 17, 227–36.

[13] Silva, AM, da; Schneider, VC; Pereira, CAM. Propriedades químicas e farmacológicas do licopeno [Chemical and pharmacological properties of lycopene]. *Rev Eletr Farm*, 2009, 6.

[14] Boileau, TWM; Boileau, AC; Erdman, JW. Bioavailability of all-trans and cis-isomers of lycopene. *Exp Biol Med (Maywood)*, 2002, 227, 914–9.

[15] Willcox, JK; Catignani, GL; Lazarus, S. Tomatoes and Cardiovascular Health. *Crit Rev Food Sci Nutr*, 2003, 43, 1–18.

[16] Boileau, AC; Merchen, NR; Wasson, K; Atkinson, CA; Erdman, JW, Jr. Cis-Lycopene Is More Bioavailable Than Trans-Lycopene

In Vitro and *In Vivo* in Lymph-Cannulated Ferrets. *J Nutr*, 1999, 129, 1176–81.

[17] Bhardwaj, M; Kumar, A. Neuroprotective Effect of Lycopene Against PTZ-induced Kindling Seizures in Mice: Possible Behavioural, Biochemical and Mitochondrial Dysfunction. *Phytother Res*, 2016, 30, 306–13. https: //doi.org/10.1002/ptr.5533.

[18] Chen, P; Xu, S; Qu, J. Lycopene Protects Keratinocytes Against UVB Radiation-Induced Carcinogenesis via Negative Regulation of FOXO3a Through the mTORC2/AKT Signaling Pathway. *J Cell Biochem*, 2018, 119, 366–77.

[19] Durairajanayagam, D; Agarwal, A; Ong, C; Prashast, P. Lycopene and male infertility. *Asian J Androl*, 2014, 16, 420–5.

[20] Ahuja, KDK; Pittaway, JK; Ball, MJ. Effects of olive oil and tomato lycopene combination on serum lycopene, lipid profile, and lipid oxidation. *Nutrition*, 2006, 22, 259–65.

[21] Rattanavipanon, W; Nithiphongwarakul, C; Sirisuwansith, P; Chaiyasothi, T; Thakkinstian, A; Nathisuwan, S; et al. Effect of tomato, lycopene and related products on blood pressure: A systematic review and network meta-analysis. *Phytomedicine*, 2021, 153512.

[22] Saini, RK; Rengasamy, KRR; Mahomoodally, FM; Keum, YS. Protective effects of lycopene in cancer, cardiovascular, and neurodegenerative diseases: An update on epidemiological and mechanistic perspectives. *Pharmacol Res*, 2020, 155, 104730.

[23] Xie, Z; Yang, F. The effects of lycopene supplementation on serum insulin-like growth factor 1 (IGF-1) levels and cardiovascular disease: A dose-response meta-analysis of clinical trials. *Complement Ther Med*, 2021, 56, 102632.

[24] Gajendragadkar, PR; Hubsch, A; Mäki-Petäjä, KM; Serg, M; Wilkinson, IB; Cheriyan, J. Effects of oral lycopene supplementation on vascular function in patients with cardiovascular disease and healthy volunteers: a randomised controlled trial. *PLoS One*, 2014, 9, e99070.

[25] Kohlmeier, L; Kark, JD; Gomez-Gracia, E; Martin, BC; Steck, SE; Kardinaal, AF; et al. Lycopene and myocardial infarction risk in the EURAMIC Study. *Am J Epidemiol*, 1997, 146, 618–26.

[26] Rissanen, T; Voutilainen, S; Nyyssönen, K; Salonen, R; Salonen, JT. Low plasma lycopene concentration is associated with increased intima-media thickness of the carotid artery wall. *Arterioscler Thromb Vasc Biol*, 2000, 20, 2677–81.

[27] Rissanen, T; Voutilainen, S; Nyyssönen, K; Lakka, TA; Sivenius, J; Salonen, R; et al. Low serum lycopene concentration is associated with an excess incidence of acute coronary events and stroke: the Kuopio Ischaemic Heart Disease Risk Factor Study. *Br J Nutr*, 2001, 85, 749–54.

[28] Rissanen, TH; Voutilainen, S; Nyyssönen, K; Salonen, R; Kaplan, GA; Salonen, JT. Serum lycopene concentrations and carotid atherosclerosis: the Kuopio Ischaemic Heart Disease Risk Factor Study. *Am J Clin Nutr*, 2003, 77, 133–8.

[29] Hu, MY; Li, YL; Jiang, CH; Liu, ZQ; Qu, SL; Huang, YM. Comparison of lycopene and fluvastatin effects on atherosclerosis induced by a high-fat diet in rabbits. *Nutrition*, 2008, 24, 1030–8.

[30] Ried, K; Fakler, P. Protective effect of lycopene on serum cholesterol and blood pressure: Meta-analyses of intervention trials. *Maturitas*, 2011, 68, 299–310.

[31] McEneny, J; Wade, L; Young, IS; Masson, L; Duthie, G; McGinty, A; et al. Lycopene intervention reduces inflammation and improves HDL functionality in moderately overweight middle-aged individuals. *J Nutr Biochem*, 2013, 24, 163–8.

[32] Oborna, I; Wojewodka, G; De Sanctis, JB; Fingerova, H; Svobodova, M; Brezinova, J; et al. Increased lipid peroxidation and abnormal fatty acid profiles in seminal and blood plasma of normozoospermic males from infertile couples. *Hum Reprod*, 2010, 25, 308–16.

[33] Nouri, M; Amani, R; Nasr-Esfahani, M; Tarrahi, MJ. The effects of lycopene supplement on the spermatogram and seminal

oxidative stress in infertile men: A randomized, double-blind, placebo-controlled clinical trial. *Phytother Res*, 2019, 33, 3203–11.

[34] Gupta, NP; Kumar, R. Lycopene therapy in idiopathic male infertility--a preliminary report. *Int Urol Nephrol*, 2002, 34, 369–72.

[35] Park, E. Birth weight was negatively correlated with plasma ghrelin, insulin resistance, and coenzyme Q10 levels in overweight children. *Nutr Res Pract*, 2010, 4, 311–6.

[36] Sluijs, I; Beulens, JWJ; Grobbee, DE; van der Schouw, YT. Dietary carotenoid intake is associated with lower prevalence of metabolic syndrome in middle-aged and elderly men. *J Nutr*, 2009, 139, 987–92.

[37] McMorrow, AM; Connaughton, RM; Magalhães, TR; McGillicuddy, FC; Hughes, MF; Cheishvili, D; et al. Personalized Cardio-Metabolic Responses to an Anti-Inflammatory Nutrition Intervention in Obese Adolescents: A Randomized Controlled Crossover Trial. *Mol Nutr Food Res*, 2018, 62, e1701008.

[38] Wiese, M; Bashmakov, Y; Chalyk, N; Nielsen, DS; Krych, Ł; Kot, W; et al. Prebiotic Effect of Lycopene and Dark Chocolate on Gut Microbiome with Systemic Changes in Liver Metabolism, Skeletal Muscles and Skin in Moderately Obese Persons. *Biomed Res Int*, 2019, 2019, 4625279.

[39] Li, YF; Chang, YY; Huang, HC; Wu, YC; Yang, MD; Chao, PM. Tomato juice supplementation in young women reduces inflammatory adipokine levels independently of body fat reduction. *Nutrition*, 2015, 31, 691–6.

[40] Ghavipour, M; Sotoudeh, G; Ghorbani, M. Tomato juice consumption improves blood antioxidative biomarkers in overweight and obese females. *Clin Nutr*, 2015, 34, 805–9.

[41] Negri, R; Trinchese, G; Carbone, F; Caprio, MG; Stanzione, G; di Scala, C; et al. Randomised Clinical Trial: Calorie Restriction Regimen with Tomato Juice Supplementation Ameliorates Oxidative Stress and Preserves a Proper Immune Surveillance Modulating Mitochondrial Bioenergetics of T-Lymphocytes in

Obese Children Affected by Non-Alcoholic Fatty Liver Disease (NAFLD). *J Clin Med*, 2020, 9.

[42] Zhu, R; Chen, B; Bai, Y; Miao, T; Rui, L; Zhang H; et al. Lycopene in protection against obesity and diabetes: A mechanistic review. *Pharmacol Res*, 2020, 159, 104966.

[43] Beheshti, F; Hashemzehi, M; Hosseini, M; Marefati, N; Memarpour, S. Inducible nitric oxide synthase plays a role in depression- and anxiety-like behaviors chronically induced by lipopolysaccharide in rats: Evidence from inflammation and oxidative stress. *Behav Brain Res*, 2020, 392, 112720.

[44] Song, Q; Feng, Y; Wang, L; Shen, J; Li, Y; Fan, C; et al. COX-2 inhibition rescues depression-like behaviors via suppressing glial activation, oxidative stress and neuronal apoptosis in rats. *Neuropharmacology*, 2019, 160, 107779.

[45] Zavvari, F; Nahavandi, A; Shahbazi, A. Neuroprotective effects of cerium oxide nanoparticles on experimental stress-induced depression in male rats. *J Chem Neuroanat*, 2020, 106, 101799.

[46] Niu, K; Guo, H; Kakizaki, M; Cui, Y; Ohmori-Matsuda, K; Guan, L; et al. A tomato-rich diet is related to depressive symptoms among an elderly population aged 70 years and over: a population-based, cross-sectional analysis. *J Affect Disord*, 2013, 144, 165–70.

[47] Allen, CM; Schwartz, SJ; Craft, NE; Giovannucci, EL; Groff, VLD; Clinton, SK. Changes in plasma and oral mucosal lycopene isomer concentrations in healthy adults consuming standard servings of processed tomato products. *Nutr Cancer*, 2003, 47, 48–56.

[48] Gärtner, C; Stahl, W; Sies, H. Lycopene is more bioavailable from tomato paste than from fresh tomatoes. *Am J Clin Nutr*, 1997, 66, 116–22.

[49] Di Mascio, P; Kaiser, S; Sies, H. Lycopene as the most efficient biological carotenoid singlet oxygen quencher. *Arch Biochem Biophys*, 1989, 274, 532–8.

[50] Qu, M; Zhou, Z; Chen, C; Li, M; Pei, L; Chu, F; et al. Lycopene protects against trimethyltin-induced neurotoxicity in primary cultured rat hippocampal neurons by inhibiting the mitochondrial apoptotic pathway. *Neurochem Int*, 2011, 59, 1095–103.

[51] Srivastava, S; Srivastava, AK. Lycopene; chemistry, biosynthesis, metabolism and degradation under various abiotic parameters. *J Food Sci Technol*, 2015, 52, 41–53.

[52] Khachik, F; Carvalho, L; Bernstein, PS; Muir, GJ; Zhao, DY; Katz, NB. Chemistry, distribution, and metabolism of tomato carotenoids and their impact on human health. *Exp Biol Med* (Maywood), 2002, 227, 845–51.

[53] Scheltens, P; De Strooper, B; Kivipelto, M; Holstege, H; Chételat, G; Teunissen, CE; et al. Alzheimer's disease. *The Lancet*, 2021, 397, 1577–90.

[54] Hwang, S; Lim, JW; Kim, H. Inhibitory Effect of Lycopene on Amyloid-β-Induced Apoptosis in Neuronal Cells. *Nutrients*, 2017, 9. https://doi.org/10.3390/nu9080883.

[55] Liu, CB; Wang, R; Yi, YF; Gao, Z; Chen, Y-Z. Lycopene mitigates β-amyloid induced inflammatory response and inhibits NF-κB signaling at the choroid plexus in early stages of Alzheimer's disease rats. *J Nutr Biochem*, 2018, 53, 66–71.

[56] Chen, W; Mao, L; Xing, H; Xu, L; Fu, X; Huang, L; et al. Lycopene attenuates Aβ1-42 secretion and its toxicity in human cell and Caenorhabditis elegans models of Alzheimer disease. *Neurosci Lett*, 2015, 608, 28–33.

[57] Sachdeva, AK; Chopra, K. Lycopene abrogates Aβ(1-42)-mediated neuroinflammatory cascade in an experimental model of Alzheimer's disease. *J Nutr Biochem*, 2015, 26, 736–44.

[58] Wang, J; Li, L; Wang, Z; Cui, Y; Tan, X; Yuan, T; et al. Supplementation of lycopene attenuates lipopolysaccharide-induced amyloidogenesis and cognitive impairments via mediating neuroinflammation and oxidative stress. *J Nutr Biochem*, 2018, 56, 16–25. h

[59] Yu, L; Wang, W; Pang, W; Xiao, Z; Jiang, Y; Hong, Y. Dietary Lycopene Supplementation Improves Cognitive Performances in Tau Transgenic Mice Expressing P301L Mutation via Inhibiting Oxidative Stress and Tau Hyperphosphorylation. *J Alzheimers Dis*, 2017, 57, 475–82.

[60] Polidori, MC; Mecocci, P. Plasma susceptibility to free radical-induced antioxidant consumption and lipid peroxidation is increased in very old subjects with Alzheimer disease. *J Alzheimers Dis*, 2002, 4, 517–22.

[61] Prakash, A; Kumar, A. Implicating the role of lycopene in restoration of mitochondrial enzymes and BDNF levels in β-amyloid induced Alzheimer's disease. *Eur J Pharmacol*, 2014, 741, 104–11. H.

[62] Cheng, L; Chen, L; Wei, X; Wang, Y; Ren, Z; Zeng, S; et al. NOD2 promotes dopaminergic degeneration regulated by NADPH oxidase 2 in 6-hydroxydopamine model of Parkinson's disease. *J Neuroinflammation*, 2018, 15, 243.

[63] Prema, A; Janakiraman, U; Manivasagam, T; Thenmozhi, AJ. Neuroprotective effect of lycopene against MPTP induced experimental Parkinson's disease in mice. *Neurosci Lett*, 2015, 599, 12–9.

[64] Suganuma, H; Hirano, T; Arimoto, Y; Inakuma, T. Effect of tomato intake on striatal monoamine level in a mouse model of experimental Parkinson's disease. *J Nutr Sci Vitaminol* (Tokyo), 2002, 48, 251–4.

[65] Wong, KS; Caplan, LR; Kim, JS. Stroke Mechanisms. *Frontiers of Neurology and Neuroscience*, 2016, 40, 58–71.

[66] Wu, A; Liu, R; Dai, W; Jie, Y; Yu, G; Fan, X; et al. Lycopene attenuates early brain injury and inflammation following subarachnoid hemorrhage in rats. *Int J Clin Exp Med*, 2015, 8, 14316–22.

[67] Hsiao, G; Fong, TH; Tzu, NH; Lin, KH; Chou, DS; Sheu, JR. A potent antioxidant, lycopene, affords neuroprotection against

microglia activation and focal cerebral ischemia in rats. *In Vivo*, 2004, 18, 351–6.

[68] Fujita, K; Yoshimoto, N; Kato, T; Imada, H; Matsumoto, G; Inakuma, T; et al. Lycopene inhibits ischemia/reperfusion-induced neuronal apoptosis in gerbil hippocampal tissue. *Neurochem Res*, 2013, 38, 461–9.

[69] Michinaga, S; Koyama, Y. Protection of the Blood-Brain Barrier as a Therapeutic Strategy for Brain Damage. *Biol Pharm Bull*, 2017, 40, 569–75.

[70] Schuele, SU; Lüders, HO. Intractable epilepsy: management and therapeutic alternatives. *Lancet Neurol*, 2008, 7, 514–24.

[71] Pearson-Smith, JN; Patel, M. Metabolic Dysfunction and Oxidative Stress in Epilepsy. *Int J Mol Sci*, 2017, 18.

[72] Kumar, V; Sharma, SK; Nagarajan, K; Dixit, PK. Effects of Lycopene and Sodium Valproate on Pentylenetetrazol-Induced Kindling in Mice. *Iran J Med Sci*, 2016, 41, 430–6.

[73] Chen, D; Huang, C; Chen, Z. A review for the pharmacological effect of lycopene in central nervous system disorders. *Biomed Pharmacother*, 2019, 111, 791–801.

[74] Henriques, JF; Portugal, CC; Canedo, T; Relvas, JB; Summavielle, T; Socodato, R. Microglia and alcohol meet at the crossroads: Microglia as critical modulators of alcohol neurotoxicity. *Toxicol Lett*, 2018, 283, 21–31.

[75] Zhang, K; Wang, H; Xu, M; Frank, JA; Luo, J. Role of MCP-1 and CCR2 in ethanol-induced neuroinflammation and neurodegeneration in the developing brain. *J Neuroinflammation*, 2018, 15, 197.

[76] Guest, J; Guillemin, GJ; Heng, B; Grant, R. Lycopene Pretreatment Ameliorates Acute Ethanol Induced NAD+ Depletion in Human Astroglial Cells. *Oxidative Medicine and Cellular Longevity*, 2015, 2015, e741612.

[77] Gambini, J; Gomez-Cabrera, MC; Borras, C; Valles, SL; Lopez-Grueso, R; Martinez-Bello, VE; et al. Free [NADH]/[NAD(+)]

regulates sirtuin expression. *Arch Biochem Biophys*, 2011, 512, 24–9.

[78] Lee, M; Ban, JJ; Chung, JY; Im, W; Kim, M. Amelioration of Huntington's disease phenotypes by Beta-Lapachone is associated with increases in Sirt1 expression, CREB phosphorylation and PGC-1α deacetylation. *PLoS One*, 2018, 13, e0195968.

[79] Hanahan, D; Weinberg, RA. Hallmarks of cancer: the next generation. *Cell*, 2011, 144, 646–74. https://doi.org/10.1016/j.cell.2011.02.013.

[80] Stahl, W; Sies, H. Lycopene: a biologically important carotenoid for humans? *Arch Biochem Biophys*, 1996, 336, 1–9.

[81] Song, X; Luo, Y; Ma, L; Hu, X; Simal-Gandara, J; Wang, LS; et al. Recent trends and advances in the epidemiology, synergism, and delivery system of lycopene as an anti-cancer agent. *Semin Cancer Biol*, 2021.

[82] Hasler, CM. (University of I. Functional foods: their role in disease prevention and health promotion. *Food Technol* (USA), 1998.

[83] Shahidi, F. Functional Foods: Their Role in Health Promotion and Disease Prevention. *J Food Sci*, 2004, 69, R146–9.

[84] Linnewiel-Hermoni, K; Khanin, M; Danilenko, M; Zango, G; Amosi, Y; Levy, J; et al. The anti-cancer effects of carotenoids and other phytonutrients resides in their combined activity. *Arch Biochem Biophys*, 2015, 572, 28–35.

[85] Langner, E; Lemieszek, MK; Rzeski, W. Lycopene, sulforaphane, quercetin, and curcumin applied together show improved antiproliferative potential in colon cancer cells *in vitro*. *J Food Biochem*, 2019, 43, e12802.

[86] Voskuil, DW; Vrieling, A; Korse, CM; Beijnen, JH; Bonfrer, JMG; van, Doorn, J; et al. Effects of lycopene on the insulin-like growth factor (IGF) system in premenopausal breast cancer survivors and women at high familial breast cancer risk. *Nutr Cancer*, 2008, 60, 342–53.

[87] Uppala, PT; Dissmore, T; Lau, BHS; Andacht, T; Rajaram, S. Selective inhibition of cell proliferation by lycopene in MCF-7 breast cancer cells *in vitro*: a proteomic analysis. *Phytother Res*, 2013, 27, 595–601.

[88] Preet, R; Mohapatra, P; Das, D; Satapathy, SR; Choudhuri, T; Wyatt, MD; et al. Lycopene synergistically enhances quinacrine action to inhibit Wnt-TCF signaling in breast cancer cells through APC. *Carcinogenesis*, 2013, 34, 277–86.

[89] Ngoc, NB; Lv, P; Zhao, WE. Suppressive effects of lycopene and β-carotene on the viability of the human esophageal squamous carcinoma cell line EC109. *Oncol Lett*, 2018, 15, 6727–32.

[90] Kucuk, O; Sarkar, FH; Sakr, W; Djuric, Z; Pollak, MN; Khachik, F; et al. Phase II randomized clinical trial of lycopene supplementation before radical prostatectomy. *Cancer Epidemiol Biomarkers Prev*, 2001, 10, 861–8.

[91] Graff, RE; Pettersson, A; Lis, RT; Ahearn, TU; Markt, SC; Wilson, KM; et al. Dietary lycopene intake and risk of prostate cancer defined by ERG protein expression1. *Am J Clin Nutr*, 2016, 103, 851–60.

[92] Ilic, D; Misso, M. Lycopene for the prevention and treatment of benign prostatic hyperplasia and prostate cancer: a systematic review. *Maturitas*, 2012, 72, 269–76.

[93] Schwarz, S; Obermüller-Jevic, UC; Hellmis, E; Koch, W; Jacobi, G; Biesalski, HK. Lycopene inhibits disease progression in patients with benign prostate hyperplasia. *J Nutr*, 2008, 138, 49–53.

[94] Obermüller-Jevic, UC; Olano-Martin, E; Corbacho, AM; Eiserich, JP; van der Vliet, A; Valacchi, G; et al. Lycopene inhibits the growth of normal human prostate epithelial cells *in vitro*. *J Nutr*, 2003, 133, 3356–60.

[95] Dulińska-Litewka, J; Sharoni, Y; Hałubiec, P; Łazarczyk, A; Szafrański, O; McCubrey, JA; et al. Recent Progress in Discovering the Role of Carotenoids and Their Metabolites in Prostatic Physiology and Pathology with a Focus on Prostate

Cancer—A Review—Part I: Molecular Mechanisms of Carotenoid Action. *Antioxidants*, 2021, 10, 585.

[96] Elgass, S; Cooper, A; Chopra, M. Lycopene treatment of prostate cancer cell lines inhibits adhesion and migration properties of the cells. *Int J Med Sci*, 2014, 11, 948–54.

[97] Rafi, MM; Kanakasabai, S; Reyes, MD; Bright, JJ. Lycopene modulates growth and survival associated genes in prostate cancer. *J Nutr Biochem*, 2013, 24, 1724–34.

[98] Lu, R; Dan, H; Wu, R; Meng, W; Liu, N; Jin, X; et al. Lycopene: features and potential significance in the oral cancer and precancerous lesions. *J Oral Pathol Med*, 2011, 40, 361–8.

[99] Livny, O; Kaplan, I; Reifen, R; Polak-Charcon, S; Madar, Z; Schwartz, B. Lycopene Inhibits Proliferation and Enhances Gap-Junction Communication of KB-1 Human Oral Tumor Cells. *J Nutr*, 2002, 132, 3754–9.

[100] Yang, T; Yang, X; Wang, X; Wang, Y; Song, Z. The role of tomato products and lycopene in the prevention of gastric cancer: a meta-analysis of epidemiologic studies. *Med Hypotheses*, 2013, 80, 383–8.

[101] Sahin, K; Cross, B; Sahin, N; Ciccone, K; Suleiman, S; Osunkoya, AO; et al. Lycopene in the prevention of renal cell cancer in the TSC2 mutant Eker rat model. *Arch Biochem Biophys*, 2015, 572, 36–9.

[102] Cross, B; Kazim, S; Nurhan, S; Karina, C; Osunkoya, Adeboye; Master, Viraj; et al. Lycopene in the prevention of renal cell cancer in the TSC2 mutant eker rat model. *J Urol*, 2012, 187, e63–e63.

About the Editors

Assoc. Prof. Gulcin Yildiz
Igdir University, Faculty of Engineering,
Food Engineering Department, Iğdır, Turkey

Dr. Yildiz obtained her PhD in Food Science from University of Illinois at Urbana-Champaign, USA. She is currently working at the food engineering department in Igdir University as an associate professor. She is also head of the department of gastronomy and culinary arts. She is the recipent of several prizes, awards and fellowships. She has attended several national/international scientific conferences and workshops as a speaker and shared her knowledge on food processing and engineering.

Interests: Novel food processing, power ultrasound, technological aspects (nutritional, microbial, enzymatic and chemical inactivation phenomena) in thermal and non-thermal processing studies, micro and nano-delivery systems, food drying, functional foods, fruit-vegetable technologies.

Assoc. Prof. Gökçen İzli
Bursa Technical University,
Faculty of Engineering and Natural Sciences,
Food Engineering Department, Bursa, Turkey

She is a Associate Professor in the department of food engineering at the Bursa Technical Univerity in Turkey. She has reviewed a lot of manuscripts for publication in high-impact factor journals. She has several research and review articles (SCI Journals), book chapters and books in the field of food processing and technology. She teaches several undergraduate and postgraduate classes including fruit-vegetable technology, food chemistry, advanced sensory analysis, and drying technology. She has been involved in several projects as project manager and researcher.

Interests: Fruit-vegetable technology, drying process, food processing and preservation, functional foods, food chemistry.

INDEX

α

β

A

B

C

D

E

F

G

H

I

J

L

M

Q

R

S

T

U

V

W

X

Z